Optical Scattering

Optical and Electro-Optical Engineering Series
Robert E. Fischer and Warren J. Smith, *Series Editors*

Published

ALLARD · *Fiber Optics Handbook*
NISHIHARA, HARUNA, SUHARA · *Optical Integrated Circuits*
RANCOURT · *Optical Thin Films Users' Handbook*
SMITH · *Modern Optical Engineering, Second Edition*
THELEN · *Design of Interference Coatings*

Forthcoming Volumes

FISCHER · *Optical Design*
JACOBSON · *Thin Film Deposition*
JOHNSON · *Infrared System Design*
ROSS · *Laser Communications*

Other Published Books of Interest

CSELT · *Optical Fibre Communication*
HECHT · *The Laser Guidebook*
HEWLETT-PACKARD · *Optoelectronics Applications Manual*
KAO · *Optical Fiber Systems*
KEISER · *Optical Fiber Communications*
MACLEOD · *Thin Film Optical Filters*
MARSHALL · *Free Electron Lasers*
OPTICAL SOCIETY OF AMERICA · *Handbook of Optics*
TEXAS INSTRUMENTS · *Optoelectronics*
WYATT · *Radiometric System Design*

Optical Scattering

Measurement and Analysis

John C. Stover

McGraw-Hill, Inc.

New York   St. Louis   San Francisco   Auckland   Bogotá
Caracas   Hamburg   Lisbon   London   Madrid
Mexico   Milan   Montreal   New Delhi   Paris
San Juan   São Paulo   Singapore
Sydney   Tokyo   Toronto

Library of Congress Cataloging-in-Publication Data

Stover, John C.
    Optical scattering : measurement and analysis / John C. Stover.
        p.   cm.
    ISBN 0-07-061814-3
    1. Light—Scattering.   I. Title.
QC427.4.S76   1990
535.4—dc20   90-36024
                                                            CIP

1 2 3 4 5 6 7 8 9 0   DOC/DOC   9 5 4 3 2 1 0

ISBN 0-07-061814-3

*The sponsoring editor for this book was Daniel A. Gonneau, the
editing supervisor was Alfred Bernardi, the designer was Naomi
Auerbach, and the production supervisor was Thomas G. Kowalczyk. It
was set in Century Schoolbook by McGraw-Hill's Professional and
Reference Division composition unit.*

Printed and bound by R. R. Donnelley & Sons Company.

Contents

Preface

This book originates from a set of notes developed over several years of teaching the fundamentals of light-scatter measurement and analysis to optical engineers (and those converting to optical engineering) at various conferences. Except for conference tutorials and a few isolated projects and classroom examples, very little is formally taught about the subject. The universities of Arizona, Alabama, New Mexico, and Montana State have done most of the university scatter work and combined they have probably produced less than 50 graduate students with thesis work on the subject. At the same time, as the sophistication, number, and expense of optical systems has grown over the 1970s and 1980s, optical scatter has been increasingly recognized as a serious problem. Outside the optics industry, noncontact process control and metrology scatter applications are just starting to be recognized. The high economic benefits associated with fast quality control in these higher volume industries (paper, steel, aluminum, ceramics, etc.) have created a need for new inspection techniques. Current indications are that by the year 2000, there will be more scatter metrology applications found outside the optics industry than within. As a result, engineers, with or without an optics background, are finding themselves thrust (sometimes kicking and screaming) into the position of becoming *the company scatter expert* as new applications are recognized.

Hundreds of papers have now been written on the subject using various notations, starting from different theoretical foundations and describing small facets of an increasingly complex field of study. These papers can be categorized as *scatter in theory, scatter is a system problem,* or *scatter is a metrology solution.* The intention of this book is to introduce engineers and physicists to scatter fundamentals, for *theory, problems,* and *solutions,* as well as acquaint them with the rather diverse set of background subjects and literature required to help them become *the company scatter expert.*

The first five chapters concentrate on background information. Chapter 1 is required reading for any other chapter, as it introduces much of the notation and basic concepts. Scatter is often tied to sample surface roughness and Chap. 2 overviews the various roughness terms and definitions. Scatter can be analyzed from diffraction theory as shown in Chap. 3. The fourth chapter combines the results of Chaps. 2 and 3 to convert scatter data to surface statistics. Chapter 5 discusses polarization concepts. There are some very powerful polarization techniques that can be used in various process and quality control applications. Experimental instrumentation, techniques, limitations, and problems are covered in Chap. 6. In the seventh chapter various scatter prediction techniques are presented. These include wavelength scaling for smooth optical surfaces and curve fitting for more generic samples. Chapter 8 discusses more advanced measurement and analysis techniques that take advantage of polarization for process and quality control applications. In the last chapter, scatter specifications are illustrated through the use of several examples. Each chapter indicates in its opening paragraphs what material is required for background and each chapter closes by indicating which of the following chapters contain material relating to the same topic. There are three appendices. Appendix A is a review of field theory necessary for EM wave propagation. Appendix B covers some diffraction theory calculations too detailed for Chap. 3. Appendix C contains scatter data for several different materials taken at several different wavelengths and angles of incidence. It is organized so the various plots can be looked up by either wavelength or sample material. Its purpose is to give the reader some indication of expected scatter levels that may be encountered.

John C. Stover

Acknowledgments

This book would not have been possible without the help and cooperation of a great many people. First, I have to thank my wife, Donna, and children (Sean, Shelly, and Rhys) for their patience and understanding during twenty long months of lost weekends and late suppers. I owe a tremendous debt to my coworkers at TMA. Bob Mathis and Don Bjork made it their job to lighten my load in order to provide enough time during the work week to complete the book. Marvin Bernt and Doug McGary are responsible for taking most of the scatter data that appears in the volume. I used a great deal of information generated by TMA authors for their technical publications and have had the pleasure, and advantage, of being able to discuss scatter issues with a first-class group of knowledgeable engineers and physicists who make their living doing scatter research. Dan Cheever, Kyle Klicker, Tod Schiff, and Dan Wilson, in particular, played key roles in designing and building the early instrumentation used to generate data and conclusions for the text. Michele Manry has cheerfully typed through the seemingly endless supply of Greek symbols and manuscript changes to produce the final copy. Mark Stefan did the technical drawings. If every picture is worth a thousand words, he has saved us all considerable effort. Outside of Bozeman, I am indebted to several members of the optical community for their help and support. As indicated by the book references, Dr. Gene Church is a wealth of information on profile analysis and scatter. He reviewed the entire text and took time from his schedule to discuss his views with me on many key topics. In many respects, this book could have been his to write instead of mine. Jean Bennett, Hal Bennett, Bob Breault, Tom Leonard, Steve McNeany, and Joe McNeely are just a few of the individuals who have given me the support (or needed stimulation) over the last two decades required to make the book possible. And last, but not least, I wish to thank Richard Skulski, my first industry supervisor (and good friend), for giving me the opportunity to work in this exciting technology.

Optical Scattering

Introduction to Light Scatter

This chapter discusses the origins of light scatter and various scatter sources that are commonly observed. Except for the following brief overview, the book is restricted to the measurement and analysis of scatter caused by surface, bulk, and contaminate imperfections, as opposed to scatter from individual molecules, aerosols, and resonance effects such as Raman scattering. Scatter from optically smooth components is treated as diffraction in many cases. For the special case of clean, optically smooth, reflective surfaces, there is a well-defined relationship between the scatter distribution pattern and the two-dimensional surface power spectral density (PSD) function. If the PSD is found from the scatter pattern it can be manipulated to reveal surface statistics (root-mean-square roughness, etc.) and in some cases insight may be gained into possible improvements in surface finish techniques. A simple example of this technique is given in this chapter and treated in more depth later. Scatter from windows, caused by both bulk and surface imperfections, is also introduced here and examined in more detail later. Although the mechanisms of bulk and particulate scatter do not lend themselves to the quantitative analysis used for surface scatter, they are still strong indicators of component quality and measurement of the resulting scatter patterns is a viable source of metrology. Scatter measurement is proving to be a useful inspection technique outside the optics industry as well. It can be used to detect and map component defects in a variety of materials, including painted surfaces, paper, metallic coatings, and medical implants such as artificial joints and intraocular lenses. Bulk and surface scatter may be separated through the use of special measurement techniques, so it is possible to determine whether or not surface polishing or a better material is required to reduce component scatter. This chapter introduces various sources of scatter and the analysis approaches that are described in later chapters. Chapter 6 introduces the measurement techniques needed to obtain the data used throughout the book.

1.1   The Scattering of Light

Most of the light we see is scattered light. We live in a world of objects that, with a few specular exceptions, scatter the visible spectrum diffusely. If those specular exceptions were the norm it would be a confusing existence at best. Some examples of scatter are more impressive than others. For instance, rainbows, alpine glow, sunsets, and blue sky are more awesome than the ability to read this page. These examples illustrate that, in a certain sense, we are all experienced in the observation and analysis of scattered light.

The interaction of light (electromagnetic radiation) with matter can be viewed through the classical mechanism of polarization. The charged particles (electrons and protons) associated with the atoms and molecules that compose a gas, liquid, or solid are stretched to form dipoles under the influence of an electromagnetic field. Since each atomic charge interacts with every other charge (to at least some degree), the number of dipole combinations is enormous. When dipoles are created, and/or stretched, by the electric field, energy is absorbed from the exciting field. The absorbed energy takes the form of a secondary field because accelerated charges produce electromagnetic radiation. These secondary fields do not necessarily propagate in the same direction or with the same phase as the initiating field. In some cases, part of the energy is lost to heat, causing the effect of absorption. Although in a low-pressure gas the interaction of a single molecule is nearly independent of its neighbors, for liquids and solids the situation is vastly more complicated. The interactions are not independent because the induced dipole fields associated with each molecule, and groups of molecules affect neighboring dipoles. For objects that are large compared to a wavelength, the result is further complicated by the fact that the amplitude and phase of the exciting field change as a function of material position. This complex situation defies a complete analytical description.

Rayleigh first studied scatter (1871) by considering the simple case of well-separated particles much smaller than a wavelength. His work included the determination that the scattered intensity from isolated particles which are small compared to a wavelength, is proportional to 1 over the wavelength to the fourth power. This relationship was used to explain the blue color of the sky and red sunsets. Following work eventually led to an explanation of atmospheric polarization effects as well. Mie theory (1908), after Gustov Mie, is the term often used for the mathematical solution for scatter from a sphere of both arbitrary radius and index of refraction. For particles much larger than a wavelength, shape is also a factor in the resulting scatter pattern, so Mie calculations do not provide an exact solution for many practical situations.

Because of the complexity of the situation, it is common to characterize larger scattering bodies by various macroscopic quantities. Reflectivity, transmissivity, and index of refraction are material constants that are actually the result of averaging millions of coupled scatter events. As such, the so-called *laws* of reflection and refraction are merely statistical results that are true only in an average sense and depend heavily on material homogeneity. A reflected (or transmitted) beam of light is the summation of a huge number of scatter components that are similar in direction, phase, and frequency. In this sense, scatter (or diffraction) out of the specular beam can be viewed as the result of fluctuations in an otherwise homogeneous material. If the fluctuations are periodic, then so is the scatter (as in the example of the next section), while random fluctuations produce a random scatter pattern. It is exactly this property that makes scatter measurement such a valuable tool for characterizing component quality and locating defects. This book contains more engineering than physics and is intended to explore and review the various measurement and analysis techniques available and not to explain the basic interactions of light and matter. Even so, it is well worth remembering the fragile microscopic mechanisms that paint our macroscopic view of the world.

1.2   Scatter from a Smooth Sinusoidal Surface

This section examines the special case of scatter (diffraction) from a smooth, clean, reflective sinusoidal surface. The objective is to define terms and illustrate a few concepts that will be used throughout the remainder of the book. The term *smooth* implies that surface height variations are small when compared to the wavelength of light. This assumption is almost always true for optics and has the added attraction that the required math is much easier. If you can see your face in the sample, it is optically smooth at visible wavelengths. The adjectives *clean* and *reflective* imply that the sample scatter is dominated by diffraction from surface topography and not surface contamination or bulk (subsurface) defects. As will be seen, these two assumptions are not always true and are more difficult to check.

In Chap. 4, surface statistics for arbitrary smooth, clean, reflective optics will be found by considering the sample topography to be composed of a summation of sinusoidal surfaces (through Fourier analysis), so the example of diffraction from a sinusoidal grating provides a great deal of insight into the problem of converting scattered light into surface roughness measurements. Figure 1.1 gives the geometry of the situation. The sample face is oriented perpendicular to the page (the x,y plane) with the incident light in the x,z plane at angle θ_i. This orientation causes the x,z plane to be the plane of incidence defined by the incident beam (P_i) and the specular reflection (P_0). The sinusoidal grooves on the surface are rotated parallel to the Y axis which causes all of the diffracted orders to also lie in the plane of incidence (denoted by P_n where $n = \pm 1, \pm 2$, etc.). The positions of the diffracted orders are given by the well-known grating equation

$$\sin \theta_n = \sin \theta_i + n f_1 \lambda \tag{1.1}$$

The quantity λ is the wavelength of the incident light and f_1 is the grating frequency which has units of inverse length and consequently is often referred to as a spatial frequency. The value $= 1/f_1$ is the distance between peaks on the grating. In this orientation the grating surface is described by

$$z(x, y) = a \sin (2\pi f_1 x + \alpha) \tag{1.2}$$

where a is the grating amplitude and α is an arbitrary phase that describes the location of the grating relative to $x = 0$. Notice that the location of the diffracted orders is dependent on the grating-line spacing and the light wavelength, but not on grating-line depth or light power.

The powers P_n depend on grating amplitude and are found through the use of diffraction theory. Exact solutions are available for simple

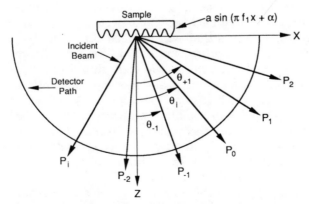

Figure 1.1   Diffraction from a sinusoidal grating.

situations; however, more complicated surfaces often require the use of approximations and as a result, several different expressions are sometimes available to describe the same situation. Solutions may be divided into the classes of scalar and vector calculations which respectively ignore and include effects of light polarization. Most scalar diffraction derivations result in diffracted orders P_n that are proportional to a summation of squared Bessel functions (Beckmann, 1963; Goodman, 1968). For normal incidence and low-angle scatter most of these relationships reduce to the proportionality shown in Eq. (1.3), where the J_n are Bessel functions of the first kind, and for *smooth surfaces* the argument $(2\pi a/\lambda)$ is much less than one. Conservation of energy is easily shown for the typical small-angle scatter case of Eq. (1.3) because the sum of the squared Bessel functions over n from minus to plus infinity is unity. Conservation of energy is not as easy to demonstrate when high-angle scatter is included.

$$P_n \approx \left[J_n \left(\frac{2\pi a}{\lambda} \right) \right]^2 \qquad (1.3)$$

A more accurate vector perturbation result, developed in the radar literature (Rice, 1951; Barrick, 1970) and based on earlier diffraction calculations (Rayleigh, 1907), has been introduced to the optical scattering literature by Church (1975; Church et al., 1977, 1979). Church's papers go far beyond examining diffraction from sinusoidal gratings and actually form the basis for our current understanding of the relationship among wavelength, angle of incidence, scatter distribution, and the surface roughness of smooth, clean, reflective optics. This relationship is commonly referred to as the *Rayleigh-Rice vector perturbation theory* or the *vector theory*. The theory consists of an equation for each of the two orthogonal polarizations. In the optics literature, s (perpendicular or occasionally TE) polarization is defined as the electric-field vector perpendicular to the plane of propagation and $\hat{p}$ (parallel or occasionally TM) polarization is defined as the electric vector in the plane of propagation. The plane of propagation is formed by the direction of propagation and the sample normal. Care is required here, as the reverse definition is common in the radar literature. Under the assumption of a perfectly conducting surface, which implies that the reflectance is unity, the equations for first-order diffraction from a sinusoidal surface are

$$P_{\pm 1}/P_i = \left(\frac{2\pi a}{\lambda} \right)^2 \cos\theta_i \cos\theta_{\pm 1} \qquad (1.4)$$

for s-polarized light and

$$P_{\pm 1}/P_i = \left(\frac{2\pi a}{\lambda}\right)^2 \frac{(1 - \sin\theta_i \sin\theta_{\pm 1})^2}{\cos\theta_i \cos\theta_s} \tag{1.5}$$

for p-polarized light.

These two equations become identical as the limit $\theta_i = \theta_{\pm 1} = 0$ is approached. For diffraction close to specular, the vector results are virtually identical to the scalar equations derived by Beckmann, Goodman, and others. All of these relationships rely on the grating equation to predict the position of each order.

We will use Eq. (1.4) for s polarization to demonstrate the importance of these simple results. Notice that if θ_i, θ_1, P_0, and P_1 are measured, the quantities a and f_1 can be easily calculated.

$$f_1 = \frac{\sin\theta_i - \sin\theta_i}{\lambda} \tag{1.6}$$

$$a = \frac{\lambda}{2\pi}\left(\frac{P_i}{P_i \cos\theta_{\pm 1} \cos\theta_i}\right)^{1/2} \tag{1.7}$$

In other words, measuring the diffracted light on either side of the specular reflection very nearly allows calculation of the surface profile. Notice that the exact profile is not available as the phase α has not been found. If the absolute phase angle between P_1 and P_0 were measured, α could also be calculated. Grating interferometers (Huntley, 1980) make use of this effect to measure transverse motion; however, for the general case of an arbitrary surface (composed of many sinusoidal gratings) measurement of all the α's is impractical.

The root-mean-square (rms) roughness σ of a sinusoidal surface is $a/\sqrt{2}$ and the average surface wavelength ℓ is obviously $1/f_1$. The rms surface slope m can be shown to be $2\pi\sigma/\ell$. The implication is that surface statistics can be evaluated even if the phase information is not known. It will be shown in Chap. 4 how these parameters can be found for more complicated surfaces by evaluating the surface power spectral density (PSD) function. It is useful to introduce this function in the case of the sinusoidal grating. The PSD may be thought of as surface roughness power per unit of spatial frequency. For the case at hand, all of the roughness is at the frequency f_1, so the PSD is a pair of impulse functions as shown in Fig. 1.2a. Readers with a background in electrical engineering or communications will see the immediate parallel to displaying power spectrums of temporal waveforms. The rms roughness is the square root of the zero moment (or integral) of

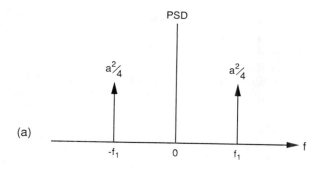

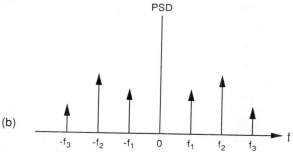

Figure 1.2 Power spectral density functions.

the PSD, and the rms slope is given by the square root of the second moment of the PSD. Because the PSD is symmetrical, it is often plotted only for $f > 0$ which is sometimes responsible for an error of $\times 2$ when the integrals are taken. This comes about because sometimes the one-sided PSDs are multiplied by $\times 2$ and sometimes they are not. If an asymmetrical PSD is computed from scatter data, it is an indication that the reflector is not a smooth, clean, and front-surface reflector and constitutes a check for these requirements. In other words, computed asymmetrical PSDs are by definition incorrect.

Figure 1.2b shows the case where three sinusoids have been summed to form a surface. This surface has a spatial bandwidth of frequencies from f_1 to f_3 and no information has been given about the relative phase of the sinusoidal components, so this PSD actually represents an infinite number of possible surface topographies all of which have the same surface statistics. The observable spatial frequency bandwidth may be caused by the sample and/or by limitations of the light-measurement instrumentation (scatterometer). As will be seen later, knowing the bandwidth limits can be critical to making valid comparisons of data taken on different instruments. Examination of the grating equation reveals that as the spatial frequency increases the diffraction angle from the specular reflection (zero order) also increases. Eventually a maximum spatial frequency is reached that diffracts along the surface of the grating (θ_s = 90°). Spatial frequencies higher than this maximum value diffract into the surface and contribute to absorption by the sample. A minimum observable spatial frequency is defined by the ability to measure close to the specular beam. This can be enhanced by using a converging beam that comes to a focus on the detector observation path. After the minimum measurable angle from the specular beam is determined, the grating equation can be used to calculate the minimum observable spatial frequency as a function of both wavelength and angle of incidence.

Another observation that can be drawn from this simple example involves the minimum required light spot size on the sample. In order to have diffraction from a grating, the spot diameter must be larger than a spatial wavelength ($1/f$). A rule of thumb is that three to five spatial wavelengths must be present in the spot to have well-defined diffraction. Thus, the spot size also places a limit on the minimum observable frequency that can be measured. Near-specular scatter measurement is an important issue for many optical imaging systems. The sinusoidal grating example can also be used to illustrate some practical measurement considerations. The tacit assumption has been made that when any of the diffracted powers are measured, the detector aperture is centered on the diffracted beam and is large enough to capture all of the power. Consider a measurement made by rotating the detector in the plane of incidence about the illuminated grating on the semicircle shown in Fig. 1.1. As the aperture approaches each diffracted spot center, the measured power increases to a maximum, holds steady, and then declines again to zero as the aperture leaves the spot. The measured width and shape of the diffracted beam is determined by aperture shape and width, as well as beam shape and width. Mathematically the measured result is known as the convolution of the beam and the aperture. Wide apertures and spots limit the degree to which closely spaced beams can be separated (or resolved) by the measurement. The situation is often improved by focusing the incident beam onto the detector path (thus reducing the spot size) and by using small detector apertures. Techniques that allow measurement to within the 0.01- to 0.1-degree region from specular for many samples are discussed in Chap. 6.

Consideration of the single frequency grating example makes it clear that once the values of a and f are known, it is possible to calculate (or predict) the scatter pattern that would result from other angles of incidence or wavelengths. These predictions can easily be made for various surfaces by using the grating equation to find scatter angles and the appropriate form of the vector perturbation theory to determine scattered powers. Obviously these predictions depend on the *smooth, clean, reflective* assumptions holding true for the new angles and wavelengths as well. The ability to scale results in wavelength can be an economically attractive alternative to taking scatter data at several wavelengths. Chapter 7 discusses this topic in more detail and gives the results of some experiments done to check predictions based on topographic scatter.

The case of diffraction from a sinusoidal grating illustrates many of the basic issues associated with scatter measurement and interpretation. It is often useful to return to this simple example in order to understand more complicated measurement situations.

1.3   Scatter from Other Surfaces

One of the major concepts presented in this book, is the one-to-one relationship between reflector-surface topography and the resulting light-scatter pattern. In fact, Chaps. 3 and 4 and part of Chap. 7 are almost solely devoted to developing the relationship between surface topography and reflective scatter. The mathematical expression of this relationship, which is founded on the Rayleigh-Rice vector perturbation theory, can be used to compute surface statistics from measured scatter patterns and is a sensitive noncontact metrology technique. However, a qualitative appreciation of the relationship between scatter and reflector topography allows insight into how a given surface (or manufacturing technique) may effect scatter, or conversely, what an observed scatter pattern implies about reflector topography. An insightful overview of this relationship can be achieved without the mathematics of the later chapters and that is the goal of the arm-waving arguments presented in this section.

The concepts are illustrated in Fig. 1.3. The geometry, shown in Fig. 1.3, consists of an illuminated reflective sample located in the x,y plane and scattering onto the x_0,y_0 observation plane. Figures 1.3a to f illustrate scatter patterns observed in x_0,y_0 for various reflector topologies. The perfectly smooth surface of Fig. 1.3a scatters no light and only the specular reflection is observed in x_0,y_0. The sinusoidal grating in Fig. 1.3b produces the scatter pattern predicted in the last section. The spacing of the diffracted orders can be calculated from the grating equation. The cusp-shaped surface of Fig. 1.3c diffracts several orders onto the observation screen. If the sinusoidal gratings that correspond to the pairs of diffraction spots are added together (with the correct phase) the result will be the cusp-shaped surface. In Chap. 3, it will be shown that the diffracted electric-field amplitudes from these sinusoidal component gratings fall off as the inverse square of

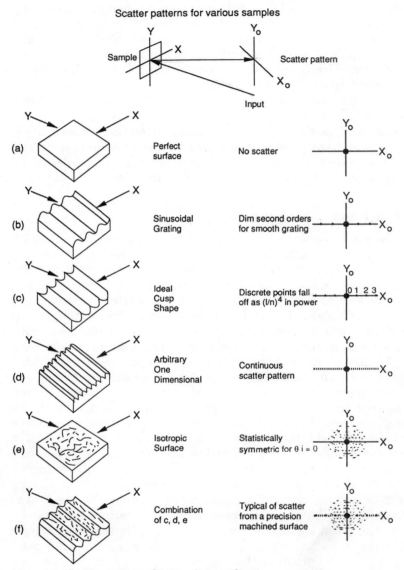

Figure 1.3 Scatter patterns from various surfaces.

the order number $(1/n^2)$. Because the diffracted light from sinusoidal gratings falls off as the square of the amplitude, the diffracted intensities of Fig. 1.3c will drop off as $1/n^4$. In Fig. 1.3d a gratinglike surface of arbitrary cross section diffracts a band of light onto the X_0 axis. Many spatial frequencies are present in this surface and each one of them produces a pair of diffraction spots. Because the band of scattered light is essentially continuous, we can infer that basically all spatial frequencies (over some bandwidth) are present in the reflector surface.

The surfaces of Figs. 1.3*b,c*, and *d* scatter light onto just the x_0 axis of the observation plane. This is because the grating lines are oriented parallel to the *y* axis in the reflector plane. If the sample is rotated about its surface normal, the scatter pattern will rotate with it according to a two-dimensional version of the grating equation presented in Sec. 3.1. Surfaces that scatter onto straight lines on the observation sphere in this manner are referred to as one-dimensional surfaces in the scatter literature. This definition does not refer to spatial dimensions (the sample does have three spatial dimensions), but to the number of spatial frequency propagation directions required to represent the surface through Fourier composition (Church et al., 1979).

In Fig. 1.3*e* a two-dimensional isotropic surface diffracts light all over the observation plane. The surface is termed two dimensional because spatial frequencies propagating in at least two directions are required to represent the surface and the resulting scatter pattern. The frequency of a sinusoidal component oriented in an arbitrary direction can be expressed as a quadrature sum of f_x and f_y components ($f^2 = f_{x^2} + f_{y^2}$). The term *isotropic* refers to the fact that if one measured the surface profile with a stylus instrument in any direction on the surface, the same average surface roughness would be found. Correspondingly, for near-normal incidence, the scatter pattern is observed to decrease from the specular peak with near-circular symmetry. Polished surfaces often exhibit near-isotropic properties. By definition, one-dimensional surfaces cannot be isotropic. The two-dimensional surface of Fig. 1.3*f* is formed by the superposition of the previous three examples. And the resulting scatter pattern is also the superposition of the previous three. This surface is similar in nature to a precision-machined (or diamond-turned) mirror. The cusp shape represents residual tool marks, the arbitrary one-dimensional roughness is caused by machine chatter and chip drag, and the background isotropic roughness is caused by random events (grain boundaries, surface scratches, and digs, etc.) that are unrelated to the periodic nature of the manufacturing process. Obviously, a great deal of information about the microscopic surface is readily available in the scatter pattern. For example, tool feed rate may be checked by analyzing the scatter pattern with the grating equation. The diffraction peaks often decrease more slowly than the expected $1/n^4$ falloff, which will be shown to indicate the presence of additional high-frequency roughness. This often takes the form of a burr on each tool mark left by the machining process. Tool wear, chip-drag, and material properties are all reflected in the scatter pattern and can be checked before the part is even removed from the machine.

1.4 Scatter from Windows and Particulates

Scatter from transmissive optics is caused by four sources: scatter from surface topography, surface contamination, bulk index fluctuations, and bulk particulates. These sources and their general scatter characteristics are introduced individually.

Surface topography. Surface topography, in both transmission and reflection, introduces phase deviations to the wavefront that may be analyzed by diffraction theory. However, transmissive scatter, which will contain contributions from two surfaces, the bulk material, and often a multiple reflection component, is much harder to analyze than scatter from a front-surface reflector. The vector perturbation theory, which has already been referred to, can be used to define polarization characteristics for some situations. Chapters 2, 3, 4, and 5 concentrate on developing these relationships.

Surface contamination. Scatter from surface contamination is less easily characterized. Particulates that are not small compared to a wavelength, produce scatter patterns whose intensity and polarization depend on particulate size, shape, and material constants. Analysis of scatter from complex shapes is difficult (Young, 1976a and b). Even correctly determining the size and density distribution of particulates on a contaminated surface is difficult. Particulates will play a significant role in producing scatter, if the samples are allowed to become contaminated. Cleaning samples and working under clean conditions can be expected to reduce surface particulate scatter to low levels in laboratory conditions. However, scatter from optics in less-controlled environments is often dominated by particulates. This is often true for space optics and necessitates sophisticated (expensive) cleaning methods. Chapter 8 discusses a measurement technique by which contaminant scatter can be separated from surface scatter.

Bulk index fluctuations. Index fluctuations may be inherent flaws in sample material or near-surface damage layers introduced by the finishing process (Brown, 1989). Because they introduce a phase change to the transmitted beam, they may be treated in much the same manner as surface fluctuations introducing a phase change to the reflected beam. As a diffraction effect, scatter from these flaws has a dependence on spot size similar to that of surface fluctuations. As shown in

Fig. 1.4, acoustooptic radio frequency spectrometers make use of the effect by inducing it in materials using modulated acoustic waves. When a laser beam is propagated perpendicular to the induced index fluctuations, part of the beam is diffracted to angles determined by the acoustic frequencies. The result is a diffraction pattern that is a measure of frequency content in the acoustic signal. The resolution of these is in turn limited by the scatter-producing bulk flaws of the acoustic material.

Bulk particulates. Bulk-particulate flaws may be due to small bubbles, inclusions, and contamination. Scatter from these sources will be similar in nature to that from surface particulates except that it cannot be eliminated by cleaning. Scatter distribution and polarization are not easily related to defect characteristics and there is no minimum scatter angle associated with the illuminated spot size.

Bulk scatter caused by isolated particulates that are small compared to the wavelength of light is called Rayleigh scatter. In this case, the particulates may be contaminants or individual molecules. Rayleigh scatter from air molecules, which is proportional to the fourth power of the inverse wavelength, is one explanation for blue sky and red sunsets. Another explanation is scatter from small thermally induced index fluctuations, which behaves in essentially the same manner. On a per molecule basis, gases actually scatter more than either liquids or solids because of the independent nature of gas molecules. More molecular scatter is normally observed from liquids and solids because there are usually more molecules scattering. True

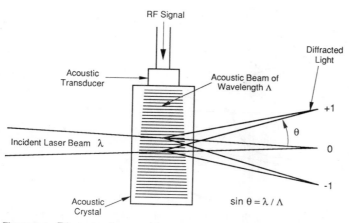

Figure 1.4   Diagram of an acoustooptic radio frequency spectrometer.

Rayleigh scatter from isolated small particles tends to be uniform in angle, but polarization dependent, as shown in Fig. 1.5. Molecular scatter from liquids and solids is dominated by forward scatter and is not uniform. However, the polarization effects illustrated in Fig. 1.5 can be exploited in instrumentation used to map bulk defects, as explained in Chap. 8. It is impossible to eliminate molecular scatter from transmissive optics.

The scatter patterns discussed in this section and the previous one are measured by sweeping a detector through the scatter field. This is usually accomplished with computer-aided instrumentation to ease the measurement process, as well as with data analysis and storage. A discussion of scatterometer instrumentation is found in Chap. 6. The next section is devoted to explaining the format commonly used to present scatter data.

1.5 Bidirectional Scatter Distribution Functions

As seen in the preceding two sections, scatter from optical components can fill the entire sphere centered about the sample. The distribution of light within the sphere is a function of incident angle, wavelength, and power, as well as sample parameters (orientation, transmittance, reflectance, absorptance, surface finish, index of refraction, bulk homogeneity, contamination, etc.). The bidirectional scatter distribution function (BSDF) is commonly used to describe scattered light patterns. The terms BRDF, BTDF, and BVDF, used for reflective, transmissive, and volume

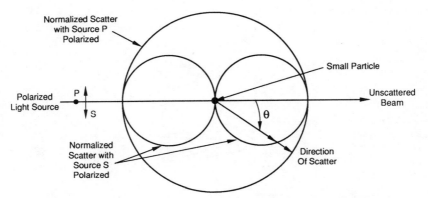

Figure 1.5 Normalized scatter intensity in the plane of the figure from a small spherical particle. The scatter pattern changes dramatically with polarization of the incident beam because the particle cannot radiate in the direction in which it is polarized.

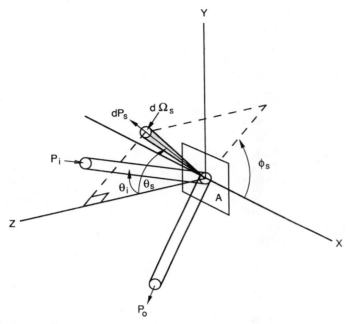

Figure 1.6   Geometry for the definition of BRDF.

samples, respectively, are merely subsets of the more generic BSDF. Although the mathematical expressions for these quantities are quite simple, they are often misunderstood. Because the BSDF is the most common form of scatter characterization and because it can be used to generate scatter specifications that enable designers, manufacturers, and users of optics to communicate and check requirements, it is well worth the minimal effort required to understand the mathematical definition and become familiar with its variations and limitations.

The derivation and notation for BRDF is credited to F. E. Nicodemus (1977) who, with his coworkers, expended considerable effort examining the problem of measuring (and defining) reflectance of optics that are neither truly diffuse nor specular (i.e., virtually all optics). The defining geometry is shown in Fig. 1.6 where the subscripts i and s are used to denote incident and scattered quantities, respectively.[1] The notation is consistent with Fig. 1.1 (where the $\theta_{\pm 1}$'s are just discrete values of θ_s) and will be used throughout the book.

---

[1]Nicodemus used r instead of s and that is still found in many papers; however, s is a more logical common subscript for both reflection and transmission and has now been adapted by many authors publishing in the scatter literature.

Nicodemus started with a fairly complicated, general case of light reflected from a surface and made several logical approximations to arrive at a simple manageable form for BRDF. Since the object here is an understanding of the use of the expression and not its complete derivation, this review will be restricted to the relatively simple case of a nearly collimated beam of light reflecting from a sample. He further simplified the situation by assuming that the beam has uniform cross section, the illuminated reflector area A is isotropic, and all scatter comes from the surface and none from the bulk. The BRDF is then defined in radiometric terms as the surface radiance divided by the incident surface irradiance. The surface irradiance is the light flux (watts) incident on the surface per unit illuminated surface area (not beam cross-sectional area). The scattered surface radiance is the light flux scattered through solid angle Ω_s per unit illuminated surface area per unit projected solid angle. The projected solid angle is the solid angle times $\cos \theta_s$. (Refer to Fig. 1.6 again.) Thus BRDF becomes

$$\text{BRDF} \equiv \frac{\text{differential radiance}}{\text{differential irradiance}} \simeq \frac{dP_s/d\Omega_s}{P_i \cos \theta_s} \simeq \frac{P_s/\Omega_s}{P_i \cos \theta_s} \qquad (1.8)$$

This equation is appropriate for all angles of incidence and all angles of scatter. Another way to look at the $\cos \theta_s$ factor is as a correction to adjust the illuminated area A to its apparent size when viewed from the scatter direction. Notice that BRDF has units of inverse steradians and depending on the relative sizes of P_s and Ω_s, can take on either very large or very small values. For example, $(P_s/P_i) \rightarrow 1$ if the entire specular reflection is measured, so (BRDF) $\rightarrow (1/\Omega_s)$ which can be very large. Scatter measurement well removed from specular generally encounters small values of P_s and requires larger apertures. The differential form is more correct and for the reasons introduced in Sec. 1.2, is only approximated when measurements are taken with a finite diameter aperture. The approximation is very good when the flux density is reasonably constant over the measuring aperture and can be very poor when using a large aperture to measure focused specular beams. The numerical value is *bidirectional* in that it depends on both the incident direction (θ_i, ϕ_i) and the scatter direction (θ_s, ϕ_s) and, as intended by Nicodemus, may be viewed as directional reflectance per unit solid angle (in steradians) of collected scatter.

The assumptions made in the Nicodemus derivation are not completely true in real measurement situations. For example, an incident laser beam is likely to have a Gaussian intensity cross section instead of one that is uniform. There is no such thing then as a truly isotropic surface and even good reflectors have some bulk scatter. And for the case of a transmitting sample, where two surfaces and the bulk are scattering, the idea of illuminated surface area becomes a little fuzzy at best. However, this only means that the BRDF, as measured, is no longer the scattered radiance divided by the incident irradiance. It still makes perfect sense to specify and measure the quantity defined in Eq. (1.8). The BSDF was defined in order to include other types of scatter, such as those from transmissive optics. The BSDF is then defined as a useful measurement parameter and not as the ratioed radiance and irradiance.

$$\text{BSDF} \equiv \frac{P_s/\Omega_s}{P_i \cos \theta_s} \tag{1.9}$$

For this reason, the $\cos \theta_s$ term is often viewed as a piece of historical baggage that no longer adds any mathematical (or physical) value to the expression. When it is dropped from the definition, the result is often called the *cosine corrected BSDF* (which is the choice made in this text) or sometimes, *the scatter function*. Care is required here as some publications will refer to the expression of Eq. (1.9) as *the BSDF* and/or *the cosine corrected BSDF*.

With the cosine dropped, the light scattered from a particular optic into any given solid angle, from any hypothetical source, may be found by multiplying the appropriate value of the cosine corrected BSDF by the incident power and solid angle. Thus, a designer with a library of typical BSDF data can address system-scatter issues and assign meaningful BSDF specifications to components. Components can then be accepted or rejected on the basis of appropriate BSDF specifications and measurements, just as interferometer measurements are used to confirm specified surface contour. Scatter specifications are covered in Chap. 9. BSDF measurements from a number of typical samples are found in App. C, which may be used as small-scatter database for systems designers.

Although scatter research and measurement facilities have been developed within the optical industry for the purpose of controlling scatter in optical systems, there are probably more scatter metrology applications in other industries. These are applications where scatter measurements are used in quality inspection or process control, as opposed to the optical industry where scatter is often a direct system problem. As will be seen in later chapters, the BSDF is still a useful format for developing specifications and communicating information outside the optics industry.

1.6 Total Integrated Scatter

The earliest scatterometers were not designed to measure BSDF, or even light scattered as a function of angle. Instead, these instruments gather (or integrate) a large fraction of the light scattered into the hemisphere in front of a reflective sample and focus it onto a single detector. The measured scattered power is then normalized by the reflected specular power and the ratio defined as the TIS, or total integrated scatter. The result is an instrument that provides repeatable results, fast sample throughput, and (without looking too closely) a single number to characterize sample scatter. TIS instrumentation introduced optical scatter as a recognized source of metrology information and these measurements are currently used as an important scatter specification. This section is devoted to a brief description of TIS measurements and its relationship to the more general scatter distribution case discussed in the previous two sections. The concepts of spatial frequencies, diffraction, and the BSDF have been introduced first because they provide valuable insights into the operation of TIS instrumentation.

In the 1950s there was considerable interest in understanding radar scatter from rough surfaces because of the problem of sea clutter associated with the detection of naval targets. A paper published in 1954 by H. Davies reported the following relationship for the fractional scattered power from a smooth, clean, conducting surface:

$$\text{TIS} \equiv \frac{P_s}{RP_i} = \frac{P_s}{P_0} \simeq \left(\frac{4\pi\sigma \cos \theta_i}{\lambda} \right)^2 \tag{1.10}$$

In addition to the smooth-surface restriction ($\lambda \ll 4\pi\sigma \cos \theta_i$), Davies assumed that the height distribution was Gaussian and that most of the light was restricted to scatter angles close to specular ($\theta_s \simeq \theta_i$). Davies extended his results to very rough surfaces and compared them to experimental data obtained at radar frequencies with encouraging results.

In 1961, H. E. Bennett and J. O. Porteus of China Lake Naval Weapons Center, published a paper that defined TIS, described the first TIS instrument, and made use of Davies smooth-surface scatter derivation. The general form of these instruments is shown in Fig. 1.7. Light strikes the sample at normal incidence and is reflected back to a detector used to measure P_0. The scattered light is gathered by a nearly hemispherical mirror (sometimes called a Coblentz sphere) that is oriented so that its center is midway between the illuminated spot on the sample and a small nearby detector. Scattered light is gathered over the region from the mirror entrance/exit aperture out to the mirror waist and is focused on the scatter detector. The mirror aperture is typically 2 to 6 degrees in diameter and the rim about 70 to 85 degrees from the sample normal. A small computer or microprocessor is often installed to calculate average TIS values. These instru-

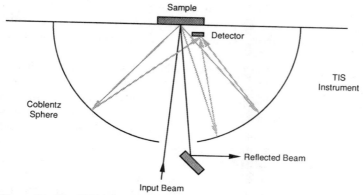

Figure 1.7 The TIS scatterometer.

ments have been built to accommodate transmissive samples and various angles of incidence, and operate at several laser wavelengths. The ability to xy raster scan the sample and plot TIS as a function of sample position has also been implemented. An alternate approach to the design of TIS instruments is found in Chap. 6. As implied by the equation above, Davies' result can be used to convert the measured TIS to the surface rms roughness.

It is worthwhile to evaluate TIS instrumentation in terms of the scatter picture introduced by the sinusoidal grating example. Davies' assumptions of a smooth, clean, conductive, Gaussian surface are more restrictive than the *smooth, clean, reflective* requirements of Sec. 1.2. Since surface profiles can be decomposed into an infinite summation of sinusoids, it is reasonable to use the results of Sec. 1.2 to analyze TIS behavior. These results are somewhat incomplete for TIS analysis in this form because Sec. 1.2 does not account for scatter out of the plane of incidence, but several insightful observations can still be made. First, notice that the hemispherical mirror-entrance aperture and rim define minimum and maximum scatter angles, respectively, and hence, minimum and maximum spatial frequency values from which scatter (diffraction) can be measured. So TIS is not truly a *total* integrated scatter measurement, but instead a measure of scatter associated with only a bandwidth of spatial frequencies. Since it has been common practice to give only the TIS value (or the corresponding rms roughness) this has complicated attempts to compare results between various labs and instruments which often operate over different spatial bandwidths. Close comparisons cannot be obtained if the same collection angles (or correspondingly the same set of sinusoidal components) are not used. This is especially true for the near-specular limit where the scattered light is usually quite intense. TIS results are more meaningful when the limiting angles are also given and should always be reported this way. Unfortunately, the literature abounds with single-number TIS scatter and rms roughness characterizations.

Large differences are often the result when rms roughness calculations, obtained by TIS, are compared to measurements on the same sample obtained by other types of surface inspection instrumentation (interferometers, profilometers, etc.). One apparent difficulty is assuring that the *Gaussian surface statistics* assumption has been met, and in fact, this served as a convenient *scapegoat* for poor comparisons for several years. However, it was later shown [Church et al., 1977, 1979; see Sec. 4.6)] that it is unnecessary to make the Gaussian assumption in order to obtain Eq. (1.10). In fact, there are more subtle reasons for problems with comparison measurements. All surface measurement systems have spatial frequency bandwidth limits and these limits must be matched before valid comparisons of measured rms roughness can be made. The low-frequency (near-specular) limit of a TIS device can be sharply defined if the reflected beam is centered in a circular entrance/exit aperture; however, the high-frequency limit is not well defined. Two effects work to discriminate against measurement of scatter from high-frequency roughness by TIS instruments. First, Davies' analysis assumed that $\theta_s \approx \theta_i$, which is clearly not true at high scatter angles. Examination of Eq. (1.4) shows that high-frequency roughness (large θ_s) scatters less light than low-frequency roughness of the same amplitude, which is not accounted for in the TIS expression for rms roughness. And second, because signal light reflected by the scatter detector (and hence not detected) goes up with angle of incidence, the detector itself discriminates against high-angle scatter. Fortunately, a majority of samples scatter most of their light close to specular so the high-angle limitations do not often pose a serious problem. TIS analysis also suffers from the fact that Davies' scalar result does not include the polarization differences apparent between Eqs. (1.4) and (1.5). Another problem with comparison measurements is that TIS, with the above reservations considered, is a true measure of area (or two-dimensional) roughness. That is, sinusoidal frequency components propagating in any (x,y) direction on the surface will scatter light to the detector. This is not true for many other measurement systems. For example, interferometers and profilometers are insensitive to roughness components that propagate perpendicular to their sampling directions. Similar difficulties arise when TIS and BRDF measurements are compared (Stover, 1984*b*). These relationships will be explored further in Chap. 4, where the calculation of surface statistics from BRDF data is discussed.

1.7 Summary

Optical scatter is a result of interaction, at a very basic level, between electromagnetic radiation and matter. Except for a few special cases, analytical solutions do not exist that completely describe the scatter pattern in terms of an input beam and the scattering element. A reasonably accurate description for scatter from a sinusoidal grating can be used to gain insight into the general principles that govern scatter distributions generated by reflective surfaces. Grating frequency determines the angular position of scatter components, while grating amplitude and frequency determine scattered power. Considering reflectors of arbitrary surface topography to be composed of a summation of sinusoidal gratings (a Fourier spectrum) will allow these results to be applied to more practical situations involving clean, smooth, reflectors. Scatter from bulk imperfections and particulates is not as easily analyzed, but can still be usefully measured and specified. Scatter patterns are conventionally presented in the form of the bidirectional scatter distribution function (BSDF). A clear understanding of this form of presentation and the limitations imposed on it by the measurement process allows its use as a standard scatter specification by designers, vendors, and users of optical systems and components. TIS measurements are fast, repeatable forms of scatter metrology that have found many applications and initiated scatter measurement as a recognized form of inspection. The sinusoidal grating example can be applied to understanding measurement results and limitations of TIS instrumentation. The underlying principles developed to reduce scatter in optical systems provide a useful background for extension of these techniques into the areas of quality inspection and process control.

This chapter has presented basic concepts and definitions, and an outline of the conversion technique used to relate scatter to surface roughness parameters. Chapters 2, 3, and 4 define surface roughness and discuss its relationship to topographic scatter via diffraction theory. Polarization of scattered light is discussed in Chap. 5. A discussion of instrumentation and measurement issues is found in Chap. 6. The seventh chapter deals with various analysis issues. For example, can you use scatter data taken in the visible to predict infrared scatter? In Chap. 8, the discussion of scatter from transmissive optics, subsurface defects, and contamination reveals that when the smooth, clean, reflective restrictions are lifted, the analysis is far more complicated and in most cases impractical. Measurements and defect detection, not calculated sample parameters, become the primary result. Polarization effects and subsurface (or bulk) scatter measurements are also discussed. The generation of scatter specifications (which is viewed as a bottom-line issue) is illustrated through the use of several practical examples in Chap. 9.

2

Surface Roughness

Everyone knows what is meant by surface roughness, or topography, and it is generally recognized that when even the smoothest surfaces are viewed in enough detail, they will exhibit some form of texture. But describing surface topography in measurable, quantitative terms is more difficult. Even the simple surfaces of Sec. 1.3 are not easily compared for relative roughness. Are any of those surfaces inherently rougher than the others, or are they just different? This chapter reviews some of the common methods of roughness measurement and presents definitions of the terms (rms roughness, power spectral density, autocorrelation length, etc.) used to quantify surface topography. It is left to the following chapters to develop the relationship of these statistical parameters to their corresponding scatter patterns.

2.1 Profile Characterization

A real three-dimensional surface, described by height z over an x,y plane requires a huge amount of information to completely describe it. Given two such complete descriptions, how does one decide which is rougher? What sort of measurable, and easily reportable, quantities should be specified to characterize surface texture? These problems were first faced in the manufacture of machined parts. One of the early methods used a set of roughness standards that were compared to machined parts by scraping the two surfaces with a thumbnail. This was followed with stylus devices that operated something like a phonograph needle on a record. As the stylus is moved across the surface, its vertical motion is converted to an electrical signal which is plotted to give an indication of surface profile. In order to interpret the resulting profile information, the signal can be processed to give the average height deviation from the surface mean. Thus, the complex profile information is converted to one number which is easy to understand and compare. This section gives the mathematical definitions of the various statistical quantities that are often used to quantify surface profiles. Additional roughness parameters can be defined, calculated (Dagnall, 1980; Thomas, 1982), and even standardized (ANSI/ASME B46.1-1985), but are not of direct interest in exploring the relationship between smooth surface topography and light scatter.

The more commonly used roughness parameters are most easily introduced in terms of a one-dimensional surface (or profile) $z(x)$. The average, mean, or expected value of $z(x)$ over distance L is denoted as $\bar{z}$ and is defined as follows.

$$\bar{z} = \lim_{L \to \infty} \frac{1}{L} \int_0^L z(x)\, dx \qquad (2.1)$$

The surface $z(x) = \bar{z}$ would be considered perfectly smooth. Roughness is defined in terms of deviations from the mean value. The arithmetic average roughness, σ_a (sometimes a.a. or R_a) is given by

$$\sigma_a = \lim_{L \to \infty} \frac{1}{L} \int_0^L |z(x) - \bar{z}|\, dx \qquad (2.2)$$

This definition of roughness became a standard within the machine-tool industry because it is obtained naturally from stylus measurements. Early stylus instruments made use of the fact that the average vertical velocity of a probe tracing the surface at constant horizontal velocity is very nearly the arithmetic average roughness. The averaging was conveniently accomplished by using a probe whose vertical position was proportional to the probe transducer voltage. The electrical signal was applied to a dc meter to accomplish averaging. The result is nearly exact for sinusoidal surfaces and poor for surfaces composed of long flat sections interrupted by sudden jumps. Thus a straightforward, repeatable method was developed to quantify roughness. More sophisticated profilometers now digitize the measured profile and use a computer to perform the required statistical analysis. Stylus-generated profiles tend to discriminate against high-frequency roughness in a manner that is not easy to compensate for in postprocessing (Wilson, 1987; Church and Takacs, 1988; Church et al., 1988).

Another surface-height average is the root-mean-square (rms) roughness σ (or R_q in machining terminology). Optical surface roughness measurements have taken advantage of the fact that the rms roughness can be obtained directly from scattered light measurements. This was initiated in the early 1960s, as mentioned in Sec. 1.6, by exploiting the convenient relationship between σ and total integrated scatter (TIS). The rms roughness is also defined in terms of surface-height deviations from the mean surface.

$$\sigma = \left\{ \lim_{L \to \infty} \frac{1}{L} \int_0^L [z(x) - \bar{z}]^2 \, dx \right\}^{1/2} \tag{2.3}$$

The name, root mean square, is obtained from the mathematical operations used in the equation. This definition depends upon the existence of the limit, as L approaches infinity, which is satisfied for real surfaces [but not for all functions $z(x)$]. Notice that the definitions of both a.a. and rms roughness are independent of adding or subtracting a constant (or dc) value to the surface function $z(x)$.

The values of either average are obtained by direct substitution into the equations. For example, if $z(x) = a \sin(2\pi f x)$, as in Sec. 1.2 then $\bar{z} = 0$ and the two roughness averages may be evaluated as follows:

$$\sigma_a = \frac{2f}{M + \Delta} \int_0^{\frac{M + \Delta}{2f}} \left| a \sin(2\pi f x) \right| dx \simeq \frac{2a}{\pi} \left(1 + \frac{1 - 2\Delta - \cos(\pi\Delta)}{2M} \right) \tag{2.4}$$

$$\sigma = \left[\frac{2f}{M + \Delta} \int_0^{\frac{M + \Delta}{2f}} a^2 \sin^2(2\pi f x) \, dx \right]^{1/2} \simeq \frac{a}{\sqrt{2}} \left[1 - \frac{\sin(2\pi\Delta)}{4M\pi} \right] \tag{2.5}$$

In the above equations the integration has been evaluated over an integer number of half-surface wavelengths (M) plus a fractional half wavelength (Δ). The fraction results in small deviations from the values expected for sinusoids. These deviations approach zero as the limit of integration approaches infinity. The approximation sign results from ignoring terms in inverse M^2. In practice, it is merely necessary to assure that the integration limits are large enough so that calculated representative surface-height averages are not dominated by integration over fractional wavelengths. If M is 2 or larger, then the maximum errors may be found as about -5 percent in σ_a and $+4$ percent in σ from Eqs. (2.4) and (2.5).

For most physically realizable surfaces, the two representative surface heights, σ and σ_a, are usually pretty close, and they are identical for the unrealizable square-wave surface, as shown in the following table:

TABLE 2.1 Comparison of Root-Mean-Square and Arithmetic-Average Roughness for Various Surface Profiles. All of the Profiles are Defined with Peak-to-Valley Height of 2a and Zero Mean. Evaluation is Over an Integer Number of Wavelengths.

Waveform	σ_a	σ
Triangle	$0.5a$	$\dfrac{a}{\sqrt{3}} \approx 0.58a$
Sinusoid	$\dfrac{2a}{\pi} \approx 0.64a$	$\dfrac{a}{\sqrt{2}} \approx 0.71a$
Square wave	a	a
Circular cusp	$0.51a$	$0.60a$

Another surface of interest is the periodic, cusp-shaped surface, shown in Fig. 2.1, that is approached (but never realized) when a circular tipped tool is used to finish a surface. The ideal profile is similar in cross section to the optical surfaces produced by precision machining (diamond turning). Because the surface is periodic, it is again important to either limit the integral to an integer number of periods or extend the integration far enough so that a partial period does not dominate the result. In principle, almost arbitrarily smooth surfaces can be achieved if the tool radius R is kept much larger than the tool feed per revolution d. There are two approaches to evaluating the rms roughness. The obvious one is to use the relationship for $z(x)$ over 0 to $+d/2$ given in Fig. 2.1 and then evaluate $z(x)$ and σ over the integral limits 0 to $d/2$ using Eq. (2.3). After the required math, this results in

$$\sigma = 0.037 d^2/R \tag{2.6}$$

A less obvious (but less painful) approach is to use a table of Fourier transforms to express the cusp shape as a summation of sinusoids.

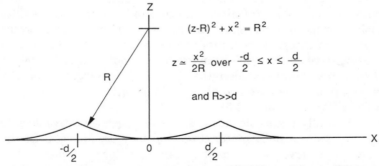

Figure 2.1 The ideal cusp-shaped surface produced by precision machining a surface with tool radius R and feed rate d.

$$z(x) = a_0/2 + \sum_{n=1}^{z} a_n \cos (2n\pi x/d)$$

where

$$a_0 = d^2/12R \quad \text{and} \quad a_n = \frac{(-1)^n d^2}{2(n\pi)^2 R}$$

The mean surface height $\bar{z}$ is the dc term in the expansion $a_0/2$ and, just as before, conveniently drops out of the integral. The phase term of each component, 0 or 180 degrees [given by $(-1)^n$] acts to alternate the signs of each term. The square on the infinite series looks messy but all of the cross-product terms have zero averages so they drop out also. All that remains is the square root of the sum of the mean square values of the individual Fourier components and these can be summed by inspection to give:

$$G = \left(\sum_{n=1}^{z} \frac{a_n^2}{2} \right)^{1/2} = \left(\frac{1}{8\pi^4} \sum_{n=1}^{z} \frac{1}{n^4} \right)^{1/2} \frac{d^2}{R} \approx 0.0373 \frac{d^2}{R} \tag{2.8}$$

The infinite sum converges to $\pi^4/90$ to obtain essentially the same result as Eq. (2.6).

This method is interesting from several other aspects as well. Notice that if only one term is evaluated, the expression for σ is the one obtained in Eq. (2.5) for a single sinusoidal grating. And when several sinusoids are added together the result is similar to that shown in Fig. 1.2b. Thus, the contributions from the individual sinusoidal components of the cusp shape add in quadrature. That is, their squares add linearly to form the mean square roughness σ^2. If a Fourier series was used that had the same amplitudes, but different phases, the same rms roughness would be found even though the components would no longer add to the cusp-shaped surface. This means that if the values of the component amplitudes are obtained without measuring the corresponding phases (via light-scatter measurements, for example) the actual surface profile cannot be obtained directly from the data, but the rms surface roughness could. For machined surfaces, the phases can be guessed to alternate by 180 degrees by examining Eq. (2.7). This has been done (Stover, 1976a) and the results were used to show the presence of a burr on the expected cusp-shaped tool marks. And finally, notice that the rms roughness does not depend at all on the component frequencies, but only on the component amplitudes a_n which are determined by d and R. The cusp profile will be analyzed further from a light-scatter viewpoint in Chap. 4.

The definitions of both σ and σ_a assume that essentially an infinite length L is available for the calculations. Clearly this is not the case.

No matter how long the sample of $z(x)$ is, there will always be a band of spatial frequencies with wavelengths so long that they cannot contribute to the calculation. There will be some frequencies that contribute only one and a fraction wavelengths, thus, errors similar to those discussed for Eqs. (2.4) and (2.5) are inherent. The trick is to get L large enough so that all the frequencies of interest are examined and, in any case, recognize and record the limitations for each practical situation. Less obvious is the limitation inherent at the high-spatial frequency (short wavelength) end of the spectrum. The defining Eqs. (2.2), (2.3), and the above examples assume that $z(x)$ is known at all points. In the real measurement world $z(x)$ is not known in equation form and instead will be sampled at discrete points. The length of the sample string defines the longest measurable spatial wavelength or largest value of L. The shortest measurable wavelength is defined by the Nyquist criteria to be twice the sample spacing (i.e., three zero crossings are required to define a full cycle). When $z(x)$ is defined as a string of N samples, given by $z_n = z(x_n)$ instead of a continuous function, the roughness values can only be estimated because of the errors associated with the bandwidth limit problems just discussed. The expressions used to provide numerical values in these situations are called *estimators* and are indicated in this text by the presence of a circumflex ^ above the estimated quantity. The commonly used estimators for the a.a. and rms roughness are given below.

$$\sigma_a \simeq \hat{\sigma}_a = \frac{1}{N} \sum_{n=0}^{N-1} |z_n - \bar{z}_n| \tag{2.9}$$

$$\sigma \simeq \hat{\sigma} = \left[\frac{1}{N} \sum_{n=0}^{N-1} (z_n - \bar{z}_n)^2 \right]^{1/2} \tag{2.10}$$

where

$$\bar{z}_n = \frac{1}{N} \sum_{n=0}^{N-1} z_n \qquad \text{and} \qquad n = 0, 1, 2, \ldots N - 1$$

The dependence of rms roughness on height alone is illustrated in Fig. 2.2. Although the two surfaces can be shown to have the same roughness, via our definitions (if appropriate sampling is used) they are likely to behave quite differently in many situations. The one on the right looks smoother and would feel smoother to the thumbnail test. Some sort of transverse quantity needs to be added to the height definitions to improve roughness characterization. Surface slope m seems a logical choice and can be defined in a manner analogous to the surface-height definitions.

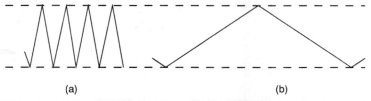

(a)   (b)

Figure 2.2   Although surfaces (a) and (b) have the same roughness values (rms or a.a.), they may behave quite differently.

$$m_a = \lim_{L \to \infty} \frac{1}{L} \int_0^L \left| \frac{dz}{dx} - \bar{z}' \right| dx \qquad \text{for arithmetic averaging} \qquad (2.11)$$

$$m = \left[\lim_{L \to \infty} \frac{1}{L} \int_0^L \left(\frac{dz}{dx} - \bar{z} \right)^2 dx \right]^{1/2} \qquad \text{for the root mean square} \qquad (2.12)$$

where

$$\bar{z}' = \lim_{L \to \infty} \frac{1}{L} \int_0^L \frac{dz}{dx} dx$$

For the case of N discrete data points representing $z(x)$ the results are again expressed as estimators.

$$m_a \simeq \hat{m}_a \frac{1}{N-1} \sum_{n=1}^{N-1} \left| \frac{z_n - z_{n-1}}{x_n - x_{n-1}} - \bar{z}n' \right| \qquad (2.13)$$

$$m \simeq \hat{m} = \left[\frac{1}{N-1} \sum_{n=1}^{N-1} \left(\frac{z_n - z_{n-1}}{x_n - x_{n-1}} - \bar{z}n' \right)^2 \right]^{1/2} \qquad (2.14)$$

where

$$\bar{z}n = \frac{1}{N-1} \sum_{n=0}^{N-1} \left(\frac{z_n - z_{n-1}}{x_n - x_{n-1}} \right) \qquad \text{and} \qquad n = 0, 1, 2, 3, \ldots N - 1$$

The surface rms height and slope may be combined to form a transverse surface length parameter or average surface wavelength ℓ.

$$\ell = 2\pi\sigma/m \qquad (2.15)$$

The following table gives the values of height, slope, and average wavelength found from these definitions for the sinusoidal grating, $z(x) = a \sin(2\pi f x)$ evaluated over an exact number of half wavelengths.

TABLE 2.2   Comparison of Sinusoidal Surface Parameters

Parameter	rms	a.a.
Height	$a/\sqrt{2}$	$2a/\pi$
Slope	$\sqrt{2}\pi fa$	$4fa$
Wavelength	$1/f$	$1/f$

Although it is reasonable to express gratinglike surfaces by one-dimensional profiles and measure (or calculate) along those profiles as discussed above, many optical surfaces need a two-dimensional expression $z(x,y)$ to adequately describe them. For these surfaces, measured roughness is often a function of measurement direction. This is illustrated by considering height or slope measurements on the sinusoidal grating of the above example. The results of Table 2.1 are obtained only if taken along the x axis. Along the y axis both the height and slope averages are zero with intermediate values obtained in other directions. Many optics will exhibit a *surface lay* and have a strong roughness dependence on direction. If the surface is isotropic, the height and slope values will be independent of measurement direction if other measurement parameters (scan length, sample interval, etc.) are fixed.

It is fairly common practice to generate *two-dimensional profiles* of surfaces by presenting several $z(x)$ traces that are offset by a small increment in y, into an isometric display of the surface. The result is a reasonable picture of surface topography, as shown in Fig. 2.3. The surface height averages can be computed from this information in a manner similar to that presented earlier. The following equations apply.

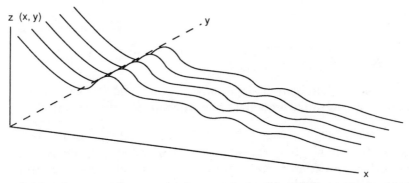

Figure 2.3   The superposition of consecutive $z(x)$ scans to form a two-dimensional surface profile.

$$\bar{z} = \lim_{\substack{L_x \to \infty \\ L_y \to \infty}} \frac{1}{L_x L_y} \int_0^{L_x} \int_0^{L_y} z(x, y)\, dy\, dx \qquad (2.16)$$

$$\sigma_a = \lim_{\substack{L_x \to \infty \\ L_y \to \infty}} \frac{1}{L_x L_y} \int_0^{L_x} \int_0^{L_y} |z(x, y) - \bar{z}|\, dy\, dx \qquad (2.17)$$

$$\sigma = \left\{ \lim_{\substack{L_x \to \infty \\ L_y \to \infty}} \frac{1}{L_x L_y} \int_0^{L_x} \int_0^{L_y} [z(x, y) - \bar{z}]^2\, dy\, dx \right\}^{1/2} \qquad (2.18)$$

To simplify the notation for the discrete sample case, z_{ni} is used to represent $z(x_n, y_i)$, where n and i are subscripts used to denote NI sample positions over the x and y directions, respectively.

$$\bar{z}_{ni} = \frac{1}{NI} \sum_{n=0}^{N-1} \sum_{i=0}^{I-1} z_{ni} \qquad (2.19)$$

$$\sigma_a \simeq \hat{\sigma}_a = \frac{1}{NI} \sum_{n=0}^{N-1} \sum_{i=0}^{I-1} |z_{ni} - \bar{z}_{ni}| \qquad (2.20)$$

$$\sigma \simeq \hat{\sigma} = \left[\frac{1}{NI} \sum_{n=0}^{N-1} \sum_{i=0}^{I-1} (z_{ni} - \bar{z}_{ni})^2 \right]^{1/2} \qquad (2.21)$$

The concept of surface slope implies a fixed direction on the surface, so the calculation of a slope on a two-dimensional surface requires some caution. If a known direction is desired, Eqs. (2.11) to (2.12) can be used by simply substituting polar coordinates (i.e., use r instead of x). If the surface is nonisotropic, then the calculated slope can be different for different directions.

The dangers of assigning numerical values to surface-roughness parameters, without also revealing the corresponding spatial frequency bandwidths, have been alluded to in Chap. 1 and will be further discussed in Chap. 4. However, it is reasonable to consider a few numbers to put the surfaces in question into proper perspective. For optics, the calculated (or measured) values of rms surface roughness do not generally exceed an upper limit of about 100 Å. A lower limit of about 1 Å, which is being approached in practice, is imposed by atomic dimensions. The range of spatial wavelengths found on optical surfaces is much larger and varies from atomic spacings up to the surface diameter. These limits are rather extreme. A surface wavelength as short as one-light wavelength (say 5000 Å) will cause normally incident light to scatter along the surface. Thus shorter spatial wavelengths do not contribute to the scatter pattern and cannot be viewed with optical microscopes. Measurement of shorter spatial wavelengths, which are not generally considered as optical roughness, requires an electron microscope or a scanning tunneling microscope. Longer surface roughness wavelengths are limited by the rather arbitrary separation of surface topography into the categories of roughness (often called *finish* or *microroughness*) and surface contour (or figure). These two merge together at surface wavelengths in the 1- to 10-mm range. So, when one hears the term *roughness* applied to optics it generally implies surface wavelengths ranging from the optical wavelength of interest up to a few millimeters. Thus optical profile slopes are generally smaller than $50/5000 = 10^{-2}$ and often smaller than 10^{-5}. Placed on such a surface you would have to walk a mile to gain (or lose?) several inches in altitude. If you have ever driven across western Kansas you have the general idea.

The surface slopes and heights can be combined to find average surface wavelengths as in the one-dimensional case; however, in order to provide additional information the surface-power spectrum and autocovariance function must be defined. The surface-power spectrum will prove to be of particular value to surface characterization by light-scatter techniques because, as has been hinted at in Chap. 1, it is central to determining the scatter pattern from smooth, clean, front surface reflectors.

2.2 The Surface Power Spectral Density and Autocovariance Functions

The statistics of random processes have been studied in several different disciplines. In engineering circles the power spectrum has been used as a powerful statistical tool for many years. For many situations it offers a very physical view of the process under study. Correlation functions, which are used for similar purposes, have seen considerably more use among mathematicians and statisticians. Although in principal the same information is available from the two functions, one of them is often more easily (or more accurately) obtained from a given data set. In the case of both surface profile data and scatter data, the surface power spectral density (PSD) is more easily found. However, because the autocovariance function approach has been used for several years and because considerable insight can be gained by studying both approaches, it is worthwhile to define the correlation function and its relationship to the power spectrum, and to indicate the methods of obtaining each of these functions from profile data.

2.2.1 The power spectral density function from the profile

The PSD may be found directly from surface profile data. This is accomplished through the use of Fourier analysis and random signal theory, the details of which are well beyond the scope of this book. Fortunately, these topics have been the subject of considerable work in the fields of communication theory and signal processing. The conversion from electrical signals of time to profiles in space is straightforward. The following is intended as a review to allow a grasp of basic essentials. An in-depth understanding can be accomplished through the study of a small fraction of the available literature (for example, Hancock, 1961; Jenkins and Watts, 1968; Bendat and Piersol, 1971 and 1986; McGillem and Cooper, 1984).

To find the PSD, $z(x)$ is first expressed in terms of its spatial frequency content by taking its Fourier transform.

$$Z(f_x) = \mathscr{F}[z(x)] = \int_{-\infty}^{\infty} z(x)\, e^{-j2\pi f_x x} \, dx \tag{2.22}$$

The integral in Eq. (2.22) acts to replace the variable x with the spatial frequency f_x propagating in the x direction. This happens in the following way. Consider $z(x)$ to be the summation of a constant $\bar{z}$ and a large (perhaps infinite) number of sinusoids of different frequencies, phases, and amplitudes. In the integral, these components are multiplied times the exponential (or phaser). At any given frequency f_x the resulting products are all periodic functions with zero mean except for two. The periodic zero-mean terms integrate to zero over the infinite length defined by the integration limits. The nonzero-mean terms are the ones formed from $\bar{z}$ and the component at frequency f. The Fourier transform of the constant $\bar{z}$ is an impulse function at $f = 0$ with area $\bar{z}$. The component at f times the phaser at f has a sinusoid squared appearance with a nonzero mean that depends on the component amplitude. So the Fourier transform acts to calculate a function giving the amplitude frequency content of $z(x)$. Because the phaser is a complex function, both amplitude and phase information are available in the result. The transform cannot be applied to any arbitrary function $z(x)$; however, physically realizable surface profiles meet the requirements for transformation. By performing the inverse transform the surface profile may be recovered.

$$z(x) = \mathcal{F}^{-1}[Z(f_x)] = \int_{-\infty}^{\infty} Z(f)\, e^{j2\pi f_x x}\, df \qquad (2.23)$$

Fourier transforms can also be defined for deterministic functions that are known over limited ranges. For example, in practical measurement cases $z(x)$ will be known only over a finite distance L and can be considered zero elsewhere. Then

$$Z(f_x,L) = \int_{-L/2}^{L/2} z(x)\, e^{-j2\pi f_x x}\, dx \qquad (2.24)$$

which, because of the limit imposed on long surface features, will drop to zero above spatial wavelengths greater than about $2L$. And conversely,

$$z(x) = \int_{-\infty}^{\infty} Z(f_x,L)\, e^{j2\pi f_x x}\, df \qquad (2.25)$$

Borrowing from electrical engineering terminology, the distance average *roughness power* of the surface may be defined as

$$P_{ave} = \lim_{L \to \infty} \frac{1}{L} \int_{-L/2}^{L/2} z^2(x)\, dx \qquad (2.26)$$

A simple argument follows in the next three equations that will help in understanding the physical meaning of the surface power spectral density function, or PSD. If Eq. (2.25) is substituted into the above expression for P_{ave}, terms are reorganized and the order of integration changed as follows:

$$P_{ave} = \lim_{L \to \infty} \frac{1}{L} \int_{-\infty}^{\infty} Z(f_x,L) \underbrace{\int_{-L/2}^{L/2} z(x)\, e^{j2\pi f x}\, dx}_{Z(-f,L)}\, df \qquad (2.27)$$

Examination of Eq. (2.24) reveals that the integral in x is $Z(-f,L)$. Then moving the limit inside the integral and defining the absolute value of a complex function in the usual manner gives

$$P_{ave} = \int_{-\infty}^{\infty} \lim_{L \to \infty} \frac{|Z(f_x,L)|^2}{L}\, df \qquad (2.28)$$

The PSD is surface height squared (or roughness power) per unit spatial frequency. If it is integrated over frequency, the result is the average roughness power obtained in Eq. (2.28). Thus the PSD of the deterministic function $z(x)$ is given by the integrand in Eq. (2.28).

$$\text{PSD} = S_1(f_x) = \lim_{L \to \infty} \frac{|Z(f_x,L)|^2}{L} \tag{2.29}$$

The subscripts 1 and x in $S_1(f_x)$ are used to indicate that the power spectrum is of a one-dimensional profile that propagates in the x direction. The phase information is lost in the process of taking the absolute value, so the surface profile cannot be recovered from the PSD. The fact that $Z(f_x,L)$ is squared makes the PSD symmetrical in f_x. The situation is slightly more complicated when $z(x)$ is one sample of a random process—as would be the case for practical surface profile measurements. The PSD is then expressed in terms of the expected value of $|Z(f,L)|^2$ (denoted by $< >$) which is formed by averaging over an ensemble of $Z(f,L)$.

$$S_1(f_x) = \lim_{L \to \infty} \frac{\langle |Z(f_x,L)|^2 \rangle}{L} = \lim_{L \to \infty} \frac{1}{L} \left| \int_{-L/2}^{L/2} z(x)\, e^{-j2\pi f_x x}\, dx \right|^2 \tag{2.30}$$

Notice that the units of the one-dimensional PSD are length to the third power. Because the PSD is symmetrical, it is fairly common to plot only the positive frequency side. Some authors include a factor of 2 in their expressions for the PSD to account for this.

Various surface statistics can be found by calculating the even moments of the PSD. (The odd moments evaluate to zero because of symmetry.) As shown below, bandwidth-limited values of the mean square surface roughness and mean square slope are easily found. The factor of 2 in front of each integral accounts for integration over only one side of the symmetrical power spectrum. The surface curvature can be obtained from the fourth moment; however, it is seldom used or specified.

$$\sigma^2 = 2 \int_{f_{min}}^{f_{max}} (2\pi f)^0\, S_1(f)\, df \tag{2.31}$$

$$m^2 = 2 \int_{f_{min}}^{f_{max}} (2\pi f)^2\, S_1(f)\, df \tag{2.32}$$

The PSD and mean square statistics are evaluated by estimators if $z(x)$ is sampled instead of fully determined (Church and Takacs, 1988). N samples z_n at x_n are taken of the profile at spacing d, where n varies from 0 to $N - 1$. This is consistent with the notation used to generate estimators for surface statistics from the profile [Eqs. (2.10) and (2.14)]. The corresponding bandwidth and interval in spatial frequency are determined by the profile sampling distance and interval.

$$0 \leq x_n = nd \leq (N - 1)d \tag{2.33}$$

$$\frac{1}{Nd} \leq f_i \frac{i}{Nd} \leq \frac{1}{2d} \tag{2.34}$$

where

$$\Delta f_i = \frac{1}{Nd} \qquad \text{and} \qquad i \text{ varies from 1 to } N/2$$

The estimator for the one-dimensional PSD, which is similar to its integral definition [Eq. (2.32)], can be formed from these elements. One other key assumption is made before generating the PSD estimator. That is that the sample profile values represent the zero-mean surface microtopography only. All effects due to electronic noise, nonlinear instrumentation, sample tilt, curvature, and so forth, are removed. This is noted in these equations by expressing the sampled values of the profile as z_{rn}, for roughness, instead of z_n. This assumption is the key to producing simple estimators that can be easily related to another means of profile characterization—the autocorrelation function. Unfortunately, these effects cannot be ignored and will eventually dictate the manner in which surface profile data need to be analyzed.

$$S_1(f_x) \simeq \hat{S}(f_i) = \frac{d}{N} \left| \sum_{n=0}^{N-1} e^{-j2\pi i n/N} z_{rn} \right|^2 K(i) \tag{2.35}$$

The quantity $K(i)$, which is equal to 1/2 at $i = 0, N/2$ and 1 elsewhere, has been defined by Church as a *bookkeeping factor* to account for end effects. In essence, the transforms are done as if the profile segment repeats itself and $K(i)$ is used to avoid doubling the contribution of end points. The mean square roughness and slope estimators follow by summing over frequency/PSD product segments.

$$\sigma^2 \simeq \hat{\sigma}^2 = 2 \sum_{i=0}^{N/2} \hat{S}_1(f_i) \, \Delta f_i = \frac{2}{Nd} \sum_{i=0}^{N/2} \hat{S}_1(f_i) \tag{2.36}$$

$$m^2 \simeq \hat{m}^2 = 2 \sum_{i=0}^{N/2} (2\pi f_i)^2 \, \hat{S}_1(f_i) \, \Delta f_i = \frac{2}{Nd} \sum_{i=0}^{N/2} (2\pi f_i)^2 \, \hat{S}_1(f_i) \tag{2.37}$$

If the analysis is extended to a two-dimensional surface $z(x,y)$ defined over a square of dimension L, then the PSD has units of length to the fourth power. The subscript 2 is used to denote the power spectra of a two-dimensional surface.

$$S_2(f_x,f_y) = \frac{1}{L^2} \left| \int\limits_{-L/2}^{L/2}\!\!\int z(x,y)\, e^{-j2\pi(f_x x\, +\, f_y y)}\, dxdy \right|^2 \tag{2.38}$$

The shape of the two-dimensional power spectrum and its relationship to the finish of various optics is studied further in Chap. 4, where the power spectrum is found from scatter data and then used to calculate surface roughness parameters. The moments of the two-dimensional PSD will be used in Chap. 4 to find surface statistics from BRDF data.

A one-dimensional spectrum, over frequencies propagating in a fixed direction, can be found from the two-dimensional spectrum by integrating over the frequency set propagating in the orthogonal direction.

$$S_1(f_x) = \int\limits_{-\infty}^{\infty} S_2(f_x,f_y)\, df_y \tag{2.39}$$

In general, $S_2(f_x,f_y)$ cannot be found from $S_1(f_x)$. The information that describes the surface in the other direction is not contained within $S_1(f_x)$. There are two obvious exceptions. If the surface is one dimensional in nature (grating-like) and f_x propagates across the surface lay, then there is no information in the y direction, and

$$S_2(f_x, f_y) = S_1(f_x)\, \delta(f_y) \tag{2.40}$$

where $\delta(f_y)$ is a Dirac Delta function.

Also, if the surface topography is isotropic, the same statistics will be obtained regardless of direction and a single profile sweep can be used as a sample representative of the whole surface. Then Eq. (2.39) can be rewritten (Church, Takacs, and Leonard, 1989) as

$$S_1(f_x) = \int\limits_{f_x}^{\infty} \frac{fS_2(f)}{\sqrt{f^2 - f_x^2}}\, df \tag{2.41}$$

where $f^2 = f_x^2 + f_y^2$. This can be solved for $S_2(f)$ as

$$S_2(f) = -\frac{1}{8\pi} \int\limits_{f}^{\infty} \frac{dS_1(f_x)}{df_x} \frac{1}{\sqrt{f_x^2 - f^2}}\, df_x \tag{2.42}$$

By having the power spectrum available, the limits of integration used to evaluate these parameters can be varied, which facilitates

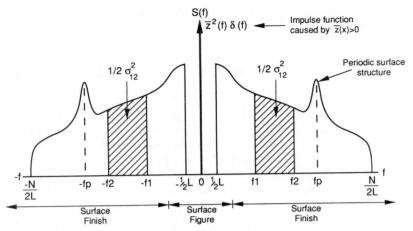

Figure 2.4 A hypothetical surface power spectrum.

comparison to other bandwidth-limited measurements. And viewing the PSD as roughness power density gives real physical intuition into the surface structure and the technique used to produce it, that is not available from the roughness averages alone. Figure 2.4 shows a hypothetical surface PSD. It covers spatial frequencies normally associated with both figure and finish. As shown here the PSD is symmetrical in frequency and has increasing roughness at increasing spatial wavelengths. Some profiles follow well-defined power relationships between roughness amplitude and frequency, known as fractals, that produce straight-line PSDs on log-log plots. The impulse function at $f = 0$ is caused by a nonzero mean value of $z(x)$. The minimum displayed frequency is due to the finite length of the profile data while the maximum frequency is limited by N the number of profile samples. The peak at f_p is caused by a periodic surface structure with spacing $1/f_p$. The integral from f_1 to f_2 allows calculation of a bandwidth-limited value of the mean square roughness σ_{12}.

The next section examines these same issues from the viewpoint of correlation functions.

2.2.2 The autocorrelation function

Correlation functions are used to study the relationship between two data sets. When the two data sets are different, the process is referred to as cross correlation. A special case, the autocorrelation function, is used to compare a data set to a translated version of itself. Autocorrelation is carried out, as shown below, by multiplying the function times the translated version of itself and then averaging. It is essentially the average of a function convolved with itself. The quantity τ is the translation and is sometimes called slip or lag. For zero translation the averaged integral is a maximum. As the translation increases and τ approaches the size of prominent surface features, the integrand will sharply reduce in average value.

$$C(\tau) = \lim_{L \to \infty} \frac{1}{L} \int_{-L/2}^{L/2} z(x)\, z(x + \tau)\, dx \qquad (2.43)$$

The autocovariance function $G(\tau)$ of $z(x)$ correlates deviations from the function mean with a translated version of itself.

$$G(\tau) = \lim_{L \to \infty} \frac{1}{L} \int_{-L/2}^{L/2} [z(x) - \bar{z}]\,[z(x + \tau) - \bar{z}]\, dx \qquad (2.44)$$

Expansion of Eq. (2.43) reveals that

$$G(\tau) = C(\tau) - \bar{z}^2 \qquad (2.45)$$

References abound on the subject (Bendat and Piersol, 1971, 1986; Bennett and Mattsson, 1989); however, caution is required as there are some variations in the literature in the definition of the term *autocovariance*. Several features are worth mentioning. The autocovariance is always an even function of τ, that is $G(\tau) = G(-\tau)$. And, as can be easily seen from its definition, its peak value at $\tau = 0$ is the surface mean square roughness. Not as obvious is the fact that its second derivative, evaluated at $\tau = 0$, is the surface mean square slope m^2.

$$G(0) = \sigma^2 = \lim_{L \to \infty} \frac{1}{L} \int_{0}^{L} [z(x) - \bar{z}]^2\, dx \qquad (2.46)$$

$$\left. \frac{d^2 G(\tau)}{d\tau^2} \right|_{\tau = 0} = m^2 \qquad (2.47)$$

The autocorrelation approach suffers from the same effective bandwidth limitations as the power spectrum did in the last section. They each use finite-length scans of the profile, sampled at finite increments, as input to equations defined for all values over an infinite scan length. Using the notation of the last section and again restricting the profile to samples of roughness only, an estimator for $C(\tau)$ can be written (Church and Takacs, 1988) as follows:

$$C(\tau) \approx \hat{C}(\tau_i) = \frac{1}{N} \sum_{n=0}^{N-1-|i|} z_{rn} z_{rn+|i|} \qquad (2.48)$$

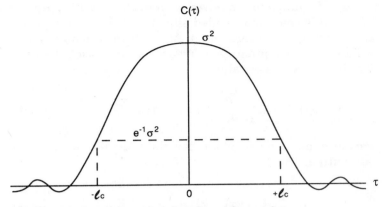

Figure 2.5   Autocovariance function of a random surface.

where $\tau_i = id$ is the slip.

The autocovariance function can take on both positive and negative values. If the surface is periodic then $G(\tau)$ will also be periodic with the same wavelength. As indicated in Fig. 2.5, the autocovariance of a surface dominated by random structure will fall from a peak at zero and eventually, as all similarity is lost between the surface and its slipped counterpart, will approach zero. The slip length required to drop from the peak value by a factor of e^{-1} is sometimes called the autocorrelation length ℓ_c. It is generally regarded as being a representative lateral dimension of surface structure, similar but not equal to the average surface wavelength ℓ defined in Eq. (2.15). Other definitions of the autocorrelation length are common. For example, it can also be evaluated by integrating the squares of either the autocorrelation function or the power spectrum as follows (Church, 1988). However, this definition yields the e^{-1} definition only when the correlation function and the power spectrum are Gaussian.

$$\ell = \frac{2}{\sigma^2} \int_0^{\infty} C^2(\tau)d\tau = \frac{1}{2\sigma^4} \int_0^{\infty} S^2(f)\, df \qquad (2.49)$$

The autocorrelation function and the power spectrum are both found from the surface profile, and both can be used to find the standard surface descriptors σ and m, and a characteristic surface length. It is reasonable to expect a relationship between the two. This is expressed in the Wiener-Khinchin relationship (Bendat, 1986) which states, in essence, that the two functions are a Fourier transform pair. Further, because $C(\tau)$ is even, only the even part of the transform phaser is required. The same is true of the inverse transform because, as has been made clear, the PSD is also an even function.

$$S_1(f) = \int_{-\infty}^{\infty} C(\tau) e^{-j2\pi f\tau} d\tau = 2 \int_{0}^{\infty} C(\tau) \cos (2\pi f\tau) d\tau \qquad (2.50)$$

$$C(\tau) = \int_{-\infty}^{\infty} S_1(f) e^{j2\pi f\tau} df = 2 \int_{0}^{\infty} S_1(f) \cos (2\pi f\tau) df \qquad (2.51)$$

When the PSD is written in terms of the autocovariance function, the impulse function located at $f = 0$ in Fig. 2.4 and discussed in the last section becomes evident.

$$S_1(f) = \int_{-\infty}^{\infty} [G(\tau) + \bar{z}^2 e^{-j2\pi f\tau}] d\tau = \int_{-\infty}^{\infty} G(\tau) e^{-j2\pi f\tau} d\tau + \bar{z}^2 \delta(f) \qquad (2.52)$$

The symmetry property leads to the use of one-sided power spectra and autocorrelation functions. As previously indicated, display and integration over only the positive frequencies is common after doubling the integrand.

Functions that are Fourier transform pairs, such as $C(\tau)$ and $S(f)$, are simply two different vehicles for expressing the same information. The PSD expresses profile statistics in spatial frequency space (units of inverse distance) and the autocorrelation function expresses the same information in slip space (units of distance). Expressing the information as a function of a variable, or the inverse variable, results in a curious relationship between the two coordinate systems. Multiplication of two functions in the variable space is equivalent to the convolution of the two functions in the inverse variable space. And the converse is also true. This fact becomes important in the next section where the effects of profile errors are discussed.

As described, the two functions are essentially equivalent in their information content. They both offer bandwidth-limited views of surface characterization. The power spectrum is more physically intuitive for most people, but some would disagree. There are some practical caveats that sometimes dictate the use of one over the other. The definition of a deterministic $z(x)$ over a finite distance L imposes a practical difficulty on using the two as a transform pair. For example, if a bandwidth-limited section of the PSD had been obtained from the profile data (or by some other technique—such as light scatter) it would be utter folly to transform to the autocorrelation function and then use that to obtain surface statistics. To do so requires that the transform be applied over frequencies from zero to infinity [Eq. (2.51)] and this cannot be accomplished because of the bandwidth limitation. Errors are introduced at no real gain in information. Just the reverse is also true. On the other hand, the estimators for $S1(f)$ and $C(\tau)$ defined for finite scans and given in Eqs. (2.35) and (2.48) transform exactly. However, there is another very basic concern that tips the balance heavily in favor of using the power spectrum to characterize surfaces—as opposed to the correlation approach. That is the topic of the next section.

2.2.3 The effects of profile measurement error

The discussion of the preceding sections has assumed that the profile data is essentially error-free. In fact this will not be the case. In addition to the bandwidth limitations imposed by sampling a finite profile length, the measuring instrument (whether profilometer or interferometer) will also have a frequency dependent response. In stylus profilometers, these types of errors are caused by stylus skip and bounce, and the nonlinear effects introduced by finite stylus radius (Wilson, 1987; Church and Takacs, 1988; Church et al., 1988). Errors vary for the different types of interferometric instruments, but include the finite pixel size of recording CCDs (charged-coupled devices), diffraction, and the usual imaging limitations of the optics employed. Electronic noise and analog filtering are also responsible for measurement limitations. Additional problems can be introduced because the exact spatial relationship (height; tilt; and roll, also called piston; slope and curvature) between the instrument and the sample is unknown. Thus, some combination of the scan length, the measurement procedure, and the instrument characteristics results in a frequency response that is not ideal. The instrument frequency response will not be a *top hat* (or flat rectangular shape) with constant value over a fixed frequency band, but will vary and is likely to change with different instrument settings.

If the resulting instrument response is known, or can be estimated, its effect can be removed, or reduced, during computer analysis of the data. This is accomplished through the use of random signal theory tools (Church and Takacs, 1986, 1988; Church et al., 1988). A common approach is to characterize the bandwidth-limited instrument frequency response and the bandwidth limitations imposed by sampling the finite scan length as functions of frequency. These can then be applied as inverse multipliers to the power spectrum found from the measured profile data. This allows estimated values of the actual power spectrum to be calculated. These corrections, which are easily expressed as functions of frequency and can be applied as multipliers in the frequency domain, would have to be transformed to functions of slip (distance) and then convolved with the autocovariance function in order to correct it for the same errors. Both of these operations require integration from minus to plus infinity and introduce bandwidth uncertainties into the corrections. Because the tools for applying the corrections needed to compensate for instrument error and bandwidth limits are more naturally (and accurately) applied in the frequency domain, the bottom line is that the power spectral density function is the logical mechanism for characterizing surface roughness and extracting surface statistics.

2.3 Summary

For a book on scatter, this chapter has spent a great deal of time on the analysis of profile data. This was done because of rampant confusion within the optical industry over the comparison of roughness statistics that are obtained by different techniques. It will soon become evident that BRDF data can be used to provide a bandwidth-limited section of the two-dimensional surface power spectrum. The definitions and discussions of this chapter are intended to act as a guide for comparing (or refusing to compare) BRDF-generated surface statistics with profile-generated surface statistics.

Roughness is commonly quantified by analyzing surface-profile data to extract various statistical averages. The surface-height deviation, from a mean value, is usually expressed as an arithmetic average in the nearly macroscopic world of the machine tool industry. In the field of optics, however, it makes sense to express roughness as a mean square because this average is proportional to scattered light measurements. In addition to surface-height averages, it also is useful to characterize roughness in terms of its average lateral dimensions. The parameters of interest are mean square slope, average spatial wavelength, and the autocorrelation length. An extremely important point is that all of these quantities are dependent, to some degree, on the measurements used to obtain them. So, although it would be nice to know that a sample has an rms roughness of, say, 50 Å with an average wavelength of 10 μm, in fact the same sample measurement done in a different direction, or with a different instrument or using a different scan length or a different sample interval, is very likely to result in different values for the roughness and wavelength. This is caused not only by instrument error and sample nonuniformity, but also by the inherent bandwidth limitations imposed on all measurements. Thus, in order to be meaningful, roughness characterization numbers should be reported with the associated measurement bandwidths.

Two approaches are commonly used to enhance surface characterization beyond that available from just two or three profile averages. These are calculation of the surface power spectral density function (PSD) and the autocovariance function. These are generated from surface-profile data and they can each be used to calculate the various profile averages of interest. However, the PSD is the preferred route because, as a function of frequency, it displays the required bandwidth limits and it can be more accurately corrected for known deviations from ideal instrument response.

Chapter 3 introduces the known relationships between surface profile and the associated reflected scatter pattern (BRDF). The PSD plays an important role as it is nearly proportional to the BRDF. In Chap. 4 relationships are presented that allow calculation of the PSD, the various profile averages, and associated bandwidth limits from the BRDF.

Scatter Calculations and Diffraction Theory

This chapter outlines the important elements of diffraction theory and gives several key results that pertain to the interpretation of measured scatter data. These results, employed in Chaps. 4 and 7, relate measured scatter from reflective surfaces to the corresponding surface roughness and consider various methods of scatter prediction. The diffraction theory results presented here and the polarization concepts found in Chap. 5 are used in Chap. 8 to outline a technique for separating surface scatter from scatter due to subsurface defects and contamination. A complete development of diffraction theory is well beyond the scope of this book; however, excellent texts on the subject are available and reference to these texts will be made in the review presented in the next four sections. The following discussions assume that the reader has some familiarity with electromagnetic field theory and the required complex math notation. Appendix A is a brief review of the elements of field theory and App. B gives details of some diffraction calculations.

3.1 Overview

When light from a point source passes through an aperture or past an edge, it expands slightly into the shadowed region. The result is that the shadow borders appear fuzzy instead of well defined. The effect is different from the one obtained by illuminating an object with an extended light source (such as the shadow of your head on this book) where the width of the reading lamp also contributes to an indistinct shadow. Well-collimated light sources (sunlight, for example) also produce fuzzy shadow edges. This bending effect, which illustrates the failure of light to travel in exactly straight lines, is called *diffraction* and is analyzed through the wave description of light.

As explained in App. A, the propagation of light is described in terms of the transverse electric field $\mathbf{E}(t,r)$, where r denotes position and t is time. The value k is $2\pi/\lambda$ and v is the light frequency. The expression in Eq. (3.1) is for a wave traveling in the direction of increasing r.

$$\mathbf{E}(t,r) = \text{Re} \left[\mathbf{e}(r) \, e^{j(kr - 2\pi vt)} \right] \tag{3.1}$$

$$\mathbf{E}(r) = \mathbf{e}(r) \, e^{jkr} \tag{3.2}$$

Phasor notation is used (the *real part* is understood) and the dependence on time, which will appear in all terms, is dropped for convenience as indicated in Eq. (3.2). The term $\mathbf{e}(r)$ gives spatial dependence. Quantities shown in **bold** are vectors and denote the polarization direction. Three common cases given below are for a plane wave traveling from $r = 0$, a spherical wave diverging from $r = 0$, and a spherical wave converging to $r = 0$. The value $\mathbf{E}_0$ is a constant in space and time. The power of the converging and diverging waves, which is proportional to $1/r^2$, follows the expected inverse square law.

$$\mathbf{E}(r) = \mathbf{E}_0 \, e^{jkr} \qquad \text{Plane wave} \tag{3.3}$$

$$\mathbf{E}(r) = \frac{\mathbf{E}_0}{r} \, e^{jkr} \qquad \text{Diverging} \tag{3.4}$$

$$\mathbf{E}(r) = \frac{\mathbf{E}_0}{r} \, e^{-jkr} \qquad \text{Converging} \tag{3.5}$$

Figure 3.1 shows the diffraction geometry for light transmitted through an aperture in the x,y plane. The aperture, centered at $r = 0$,

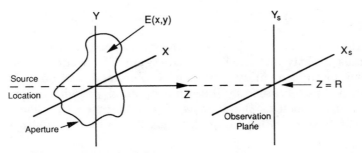

Figure 3.1 Diffraction geometry.

is typically illuminated by a point source (diverging), a collimated beam (plane wave), or a converging beam (virtual point source). In general, the aperture will modulate the transmitted light in both amplitude and phase. The modulated light leaving the aperture is given by $\mathbf{E}(x,y)$ and the object of the diffraction calculation is to find the resulting electric field $\mathbf{E}(x_s,y_s)$ in the observation plane located distance R from the aperture. The source could also be located on the $z > 0$ side of a reflective aperture (or sample).

Amplitude modulations, caused by changes in aperture reflectance or transmittance, are expressed by variations in $e(x,y)$. For example, a slit aperture changes from zero transmittance to unity, and back again, with no phase modulation. Phase modulations are caused by index of refraction changes in transmitting samples and by surface roughness on reflecting samples, and are expressed by changes in the exponential component of $\mathbf{E}(r)$.

A useful exercise is the calculation of diffracted light from a slit aperture without the benefit of using a diffraction theory result. The general nature of the solution, the approximations required, and the limitations of such an approach become immediately obvious. Consider a slit aperture of width L in the x,y plane along the y axis (perpendicular to the page) as shown in Fig. 3.2. A plane wave traveling along the z axis is incident upon the aperture and diffraction is to be observed at the x_s,y_s plane located at $z = R$. The assumption is made that $R \gg L$. Use is made of Huygens' principle, which is an intuitive statement that wave fronts can be constructed by allowing each point in a field to radiate as a spherical source. The new wave front, downstream, is then found from the envelope of the spherical fronts. Early diffraction results depended on variations of this reasoning, even though there are some obvious problems. For example, what do we do

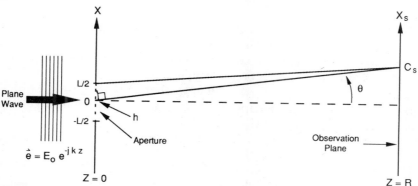

Figure 3.2   Diffraction from a rectangular aperture.

about the backward traveling wave? We will also ignore any field/ aperture interaction and assume that the field exists as presented by the source right up to the aperture edge, where it drops to zero in a sudden discontinuity. Polarization issues are ignored. This is a true *back of the envelope calculation* whose purpose is to develop insight for the more complicated issues to follow.

Two rays, leaving from $x = 0$ and $x = L/2$ and eventually interfering at coordinate C_s on the x_s axis are shown in the diagram. The path difference of the two waves is the small distance h shown in the figure. Making use of the small-angle assumptions gives

$$h = Lx_s/2R \tag{3.6}$$

At some value of x_s, h will reach the value $\lambda/2$ and the two waves will cancel in the observation plane. Within the limitations of the small-angle assumption, the same reasoning holds for all the other pairs of rays separated by $L/2$ at the aperture and reaching point C_s. Thus the condition

$$x_s = \pm n\lambda R/L \tag{3.7}$$

(where n is an integer) will result in a zero intensity value on the otherwise illuminated x_s axis. The relative intensity pattern can also be found. Ignoring polarization issues, the spherically expanding wave from a differential source, dE over dx located at x in the aperture will have an amplitude proportional to dx and inversely proportional to the distance from x. The resulting differential scalar amplitude from the differential source may be evaluated in the observation plane, where K has been used as a proportionality constant.

$$dE_s = \frac{K\,dx}{\sqrt{R^2 + (x_s - x)^2}}\, e^{jk\sqrt{R^2 + (x_s - x)^2}} \tag{3.8}$$

The approach is to integrate over x from $-L/2$ to $L/2$ and thus obtain the total field strength at C_s. In order to do the integral easily, some assumptions are made to simplify the expression for the distance r_s between x and C_s. In the amplitude component the distance is approximated as R. However in the phase component, this is inappropriate as distance errors of only half a wavelength change the sign with which a particular component is summed. For this term the radical can be expanded via the binomial theorem.

$$\sqrt{R^2 + (x_s - x)^2} = R\left[1 + \frac{(x_s - x)^2}{R^2}\right]^{1/2}$$

$$= R\left[1 + \frac{(x_s - x)^2}{2R^2} - \frac{(x_s - x)^4}{8R^4} + \cdots\right]^{1/2}$$

$$= \underbrace{R + \frac{x_s^2}{2R} - \frac{x_s x}{R}}_{\text{Fraunhofer}} + \underbrace{\frac{x^2}{2R} - \frac{x_s^4}{8R^3} \cdots}_{} \qquad (3.9)$$

Fresnel

Each additional term makes the integral more accurate and more difficult to evaluate. The two common approximations have been named after the men that made them as indicated. The larger R is the better of the approximations. This has led to the terminology *getting to the far field*, which usually implies the Fraunhofer approximation is good enough. If a source is used that converges at the observation plane a term is introduced that cancels the $x^2/2R$ term making the Fraunhofer and Fresnel approximations identical. For the example at hand we will proceed with the Fraunhofer approximation and evaluate the integral as follows:

$$E_s = \frac{KL}{R} e^{jk(R + x^2/2R)} \int_{-L/2}^{L/2} e^{-j(2\pi x_s X/\lambda R)} dx \qquad (3.10)$$

$$E_s = \frac{K}{jR} e^{jk(R + x^2/2R)} \operatorname{sinc}\left(\frac{x_s L}{\lambda R}\right) \qquad (3.11)$$

where

$$\operatorname{sinc}(x) = \frac{\sin(\pi x)}{\pi x}$$

Squaring the absolute value of the electric field and dividing by twice the impedance of free η_0 space gives the time average power density I_s (watts per unit area) as a function of x_s.

$$I_s = \frac{1}{2\eta_0}\left[\frac{KL}{R}\right]^2 \text{sinc}^2\left(\frac{x_sL}{\lambda R}\right) \tag{3.12}$$

This relationship is plotted in Fig. 3.3. Notice that the zero intensity values appear at the locations predicted earlier by Eq. (3.7). This means that the Fraunhofer approximation is equivalent to the same small-angle approximation. Patterns very much like the one in Fig. 3.3 can be observed by placing a small slit in a HeNe laser beam. The inverse aperture, a small opaque block, is easier to use (a piece of hair works just fine). The above relationships can be used to calculate the hair diameter from the diffraction pattern zeros. The proportionality constant K has not been evaluated, but this could be accomplished by integrating over the observation plane and applying the conservation of energy.

Another observation is worth making. The sinc function is the Fourier transform of the slit aperture [sometimes expressed as rect (x/L)]. In fact, Eq. (3.10) shows this explicitly. In this context, the quantity $(x_s/\lambda R)$ may be viewed as a spatial frequency, propagating in the x direction on the aperture plane. Notice that it has units of inverse length as required in Chap. 1. In fact, this is the same expression for spatial frequency that is obtained from the grating equation at normal incidence and small angles. Retracing our steps back through the development, it is easy to see that if an aperture function other than unity had been applied, the Fraunhofer approximation would result in simply taking the Fourier transform of that function. This is one of the results of scalar diffraction theory and is the basis for the field of study called Fourier optics.

Analysis of the single slit, examined above, is straightforward because the sample (the slit) is very simple. For one thing, the slit has a constant transmission amplitude across its aperture. Many samples

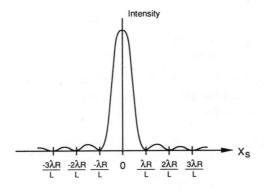

Intensity

$\frac{-3\lambda R}{L}$ $\frac{-2\lambda R}{L}$ $\frac{-\lambda R}{L}$ 0 $\frac{\lambda R}{L}$ $\frac{2\lambda R}{L}$ $\frac{3\lambda R}{L}$ X_s

Figure 3.3 Diffraction pattern from a slit.

will require that phase changes also be considered (for example, the reflective sinusoidal surface of Chap. 1). Even for the slit, the required mathematics were messy and several assumptions were needed. If the sample is somehow more complicated than the on/off nature of the slit or can be defined only statistically, the situation is far more difficult.

It has been common practice to refer to diffraction calculations as either scalar or vector, depending on whether or not polarization is considered. A more descriptive way is to label the calculations based on the mathematical approach. Most optics texts analyze diffraction by the Kirchhoff method, which is described in the next section. It can be either scalar or vector in its approach. A second method, introduced by Rayleigh in 1895, has been less well traveled because of the considerably stiffer mathematical requirements, but offers advantages in some areas. The next two sections will review these approaches.

3.2 Kirchhoff Diffraction Theory

Diffraction calculations based on the Kirchhoff theory address many of the loose ends of the last section. The proportionality constant is evaluated. Boundary conditions are handled by various approximations. The Fraunhofer and Fresnel approximations are generally treated exactly as they were in the last section. There are lots of approximations to make, leaving ample room for considerable individuality in applications to specific problems. The general scalar Kirchhoff approach is outlined in this section and a commonly used Fraunhofer equation is derived in two dimensions. This equation is then used for several easily defined cases, including the sinusoidal grating of Chap. 1.

This presentation is intended as an outline and is restricted to scalar results only. Far more complete treatments of this subject are readily available in the literature (Beckmann and Spizzichino, 1963; Goodman, 1968). The derivation is based on Green's theorem which is used to convert back and forth between volume and surface integrals over two functions. One of these functions plays the role of an unknown, to be evaluated, while the other is arbitrarily chosen. The two complex, scalar functions in question, $E(x,y,z)$ and $G(x,y,z)$, are restricted to situations where they, and their first and second derivatives, are single valued and continuous within a volume V and on its bounding surface S. In addition, since the objective here is a solution for electric-field strength of diffracted light, it is required that both functions obey the wave equation. The vector $\mathbf{n}$ is defined as the outwardly directed unit normal to the surface S. Green's theorem is expressed as

$$\int_V [G\nabla^2 E - E\nabla^2 G]\, dv = \int_S [G\nabla E - E\nabla G] \cdot \mathbf{n}ds \qquad (3.13)$$

and the wave equation is given as

$$(\nabla^2 + k^2)\mathbf{E} = (\nabla^2 + k^2)\mathbf{G} = 0 \qquad (3.14)$$

Using scalar notation, we will solve for the unknown $E(x,y,z)$ at point C_s within V and choose $G(x,y,z)$ as an arbitrary Green's function. The solution will depend not only on the choice of $G(x,y,z)$ but also on the choice of the bounding surface S surrounding V. This freedom of choice gives the solution a rather arbitrary aroma. Would a better solution have been obtained if.... But the purpose here is to follow well-trod paths.

The wave equation allows the functions to be defined throughout the volume V, and we convert to the surface evaluations using Green's theorem. The solution for $E(x,y,z)$ at C_s is then regarded as due to diffraction occurring from S. The geometry is shown in Fig. 3.4. Kirchhoff picked a unit-amplitude spherical wave expanding from the observation point C_s as his choice for $G(x,y,z)$. Therefore

$$G(r) = \frac{e^{jkr_s}}{r_s} \qquad (3.15)$$

The value r_s is the distance from C_s to the point where $G(x,y,z)$ is to be evaluated, as indicated in Fig. 3.4. The point C_s must be excluded from the volume V because of the discontinuity in $G(x,y,z)$. This is handled by creating a sphere of differential radius about C_s that is excluded from the volume. Substituting into the wave equation (to evaluate the Laplacians) and then into Green's theorem gives

$$\int_V (Gk^2E - Ek^2G)\, dV = \int_{S_s + S_a + S_0} \left[\frac{e^{jkr_s}}{r_s}\nabla E - E\nabla\left(\frac{e^{jkr_s}}{r_s}\right) \right] \cdot \mathbf{n}\, dS \qquad (3.16)$$

The four integrals are evaluated individually. The volume integral is zero by inspection. This will be true as long as there are no sources within the volume. The surface integral over S_s evaluates to $-4\pi E(x_s, y_s)$ in a straightforward manner, as the radius of the differential sphere is reduced to zero in the limit. With some difficulty, the integral over S_0 can be shown to reduce to zero for most physically realizable situations, as the radius of that surface approaches infinity. Equation (3.16) now reduces to an expression for $E(x_s, y_s)$ in terms of the field, and its normal derivative, over the infinite plane S_a directly behind the aperture.

$$E(x_s, y_s) = \frac{1}{4\pi}\int_{S_a} \left[\frac{e^{jkr_s}}{r_s}\nabla E - E\nabla\left(\frac{e^{jkr_s}}{r_s}\right) \right] \cdot \mathbf{n}\, dS \qquad (3.17)$$

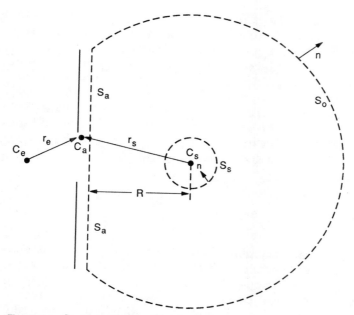

Figure 3.4 Geometry for the Kirchhoff solution.

Now, if the emission from C_e is a spherical wave given by

$$E = E_0 \frac{e^{jkr_e}}{r_e} \qquad (3.18)$$

and if the radii r_s and r_e are much larger than a wavelength then

$$\nabla\left(\frac{e^{jkr}}{r}\right) = \left(jk - \frac{1}{r}\right)\frac{e^{jkr}}{r} \simeq -jk\frac{e^{jkr}}{r} \qquad (3.19)$$

Substitution back into Eq. (3.17) gives an integral expression for $E(x_s,y_s)$ that includes contributions from the entire infinite plane behind the aperture. If the boundary conditions of the last section are imposed (i.e., outside the aperture $E = \nabla E = 0$) and inside the aperture the evaluations follow from Eqs. (3.18) and (3.19), the result is the Fresnel-Kirchhoff diffraction formula, sometimes known as the reciprocity theorem of Helmholtz.

$$E(x_s,y_s) = \frac{E_0}{j2\lambda}\int_{S_a} \frac{e^{jk(r_s + r_e)}}{r_s r_e}[\cos(\mathbf{r}_s,\mathbf{n}) - \cos(\mathbf{r}_e,\mathbf{n})]\,dS \qquad (3.20)$$

The bracketed cosines (defined in terms of the angle between the indicated vectors) are called the *obliquity factor*, which is approximately 2 for geometries where $\mathbf{r}_s$ and $\mathbf{r}_e$ are nearly perpendicular to the x,y plane. Equation (3.20) can be related to the Huygens principle in an interesting way. The integrand can be broken into two parts by factoring out the exponential, which looks like a Huygens spherical wave expanding from ds in the aperture to C_s in the observation plane. Everything that remains can be considered a complex amplitude of that spherical wave. If the observation point is moved back to the source the obliquity factor is zero (the cosines cancel) and the backward traveling wave is zero.

$$E(x_s, y_s) = \int_{S_a} [\text{complex aperture amplitude}] \frac{e^{jkr_s}}{r_s} \, dS \qquad (3.21)$$

Unfortunately there is a problem with the boundary conditions in Eq. (3.20), as it can be shown that if E and its derivative are identically zero over the aperture edge, they are also zero within the aperture. This can be addressed (Goodman, 1968) by choosing a Green's function that corresponds to a source at C_s and its mirror image, 180 degrees out of phase. This choice was motivated by the fact that G is identically zero over the aperture so that the zero value boundary condition on ∇E need not be applied.

$$G = \frac{e^{jkr_s}}{r_s} - \frac{e^{jkr_s}}{r'_s} \qquad (3.22)$$

The result is known as the Rayleigh-Sommerfield diffraction formula.

$$E(x_s, y_s) = \frac{E_0}{j\lambda} \int_{S_a} \frac{e^{jk(r_e + r_s)}}{r_e r_s} \cos(\mathbf{r}_s, \mathbf{n}) \, dS \qquad (3.23)$$

Notice that the only difference is in the obliquity factor, which is now independent of the incident angle. This solves the boundary condition problem; however, if the observation point is again moved to the source, there is a backward traveling wave. The relative merits of these two scalar representations have been studied (Wolf and Marchand, 1964) and is still a topic of interest.

Another way of looking at scalar diffraction that gives similar results and is more pleasing physically is to assume that the waves within a clear aperture progress undisturbed (no Huygens wavelets) and the diffraction pattern is caused by the superimposition of a second set of waves that originate from the aperture boundaries (Keller, 1962). Considering all of the approximations and rather arbitrary choices in approach, it is amazing that diffraction theory produces results that even resemble reality.

As long as the source is kept reasonably close to the aperture normal, our two results [Eqs. (3.20) and (3.23)] are identical except for small variations in the obliquity factor. Further, if we restrict ourselves to apertures that are small compared to r_e and r_s, the cosines are nearly $\cos \theta_i$ and $\cos \theta_s$, respectively, and are independent of x and y. Under these assumptions it is convenient to express Eq. (3.23) as

$$E(x_s, y_s) = \frac{\cos (\theta_s)}{j\lambda} \int_{S_a} E_a(x, y, 0) \frac{e^{jkr_s}}{r_s} dS \qquad (3.24)$$

where $E_a(x,y,0)$ is the incident field modulated by whatever is in the aperture. For the clear aperture studied so far,

$$E_a(x,y,0)\Big|_{\substack{Clear \\ apt.}} = \frac{E_0}{r_e} e^{jkr_e} \qquad (3.25)$$

In general $E_a(x,y,0)$ will be composed of whatever source wave impinges upon the aperture modulated in amplitude and/or phase by whatever is contained within the aperture. The radius r_s can now be evaluated much as we did in the last section to any degree of accuracy.

$$r_s = [(x_s - x)^2 + (y_s - y)^2 + R^2]^{1/2} \qquad (3.26)$$

$$r_s = R + \frac{x_s^2}{2R} + \frac{y_s^2}{2R} - \frac{x_s x}{R} - \frac{y_s y}{R} + \frac{x^2}{2R} + \frac{y^2}{2R} - \frac{x_s^4}{2R} - \frac{y_s^4}{2R} + \ldots \qquad (3.27)$$

$$\underbrace{\qquad\qquad\qquad\qquad\qquad}_{\text{Fraunhofer}}$$

$$\underbrace{\qquad\qquad\qquad\qquad\qquad\qquad\qquad\qquad}_{\text{Fresnel}}$$

Again of particular interest is the Fraunhofer approximation which gives the common scalar diffraction formula

$$E(x_s, y_s) = \frac{\cos \theta_s}{j\lambda R} e^{jk(R + x_s^2/2R + y_s^2/2R)} \int_{S_a} E_a(x,y,0) e^{-j2\pi(f_x x + f_y y)} dx \, dy \qquad (3.28)$$

The quantities f_x and f_y are defined as spatial frequencies in the aperture.

$$f_x = \frac{x_s}{\lambda R} \qquad f_y = \frac{y_s}{\lambda R} \qquad (3.29)$$

Equation (3.28) is simply the Fourier transform of the electric field within the aperture multiplied by a complex amplitude. Notice that it is identical to the one-dimensional form derived in Eq. (3.10) where the proportionality constant K has now been evaluated as $1/j\lambda$. If the obliquity factor had been left exact, the integral would have been considerably more difficult to evaluate.

Equation (3.28), with slight variations on the obliquity factor, has been used throughout the literature to evaluate a number of common diffraction problems (Goodman, 1968; Iizuka, 1985). Here it is enough to give three brief examples. First, by inspection after reviewing Eqs. (3.10) to (3.12), the diffracted intensity I_s (watts per unit area) from a rectangular aperture of dimensions L_x, L_y can be given as

$$I(p_s) = \frac{1}{2\eta_0} \left(\frac{E_0 \, L_x L_y \cos \theta_s}{\lambda R} \right)^2 \text{sinc}^2 \left(\frac{x_s L_x}{\lambda R} \right) \text{sinc}^2 \left(\frac{y_s L_y}{\lambda R} \right) \qquad (3.30)$$

where η is the medium impedance as defined in App. A.

Notice that as in the one-dimensional slit, diffracted power and the minimum/maximum diffraction locations depend on both wavelength and aperture size, and the first minima is pushed farther from specular as the aperture size decreases. The two-dimensional result shows that there is no coupling of diffraction from the x and y aperture edges.

A similar result is obtained for a circular aperture of diameter L, as shown in Eq. (3.31), where J_1 is a first-order Bessel function of the first kind and $r_s^2 = x_s^2 + y_s^2$. Known as an Airy pattern, these concentric diffraction rings are familiar to anyone who has adjusted a conventional spatial filter. The first minima does not come at quite the same distance from center as for the slit aperture.

$$I(p_s) = \frac{1}{2\eta_0} \left(\frac{E_0 k L^2 \cos \theta_s}{8R} \right)^2 \left[\frac{J_1(kLr_s/2R)}{kLr_s/2R} \right]^2 \qquad (3.31)$$

The third example, which is the sinusoidal surface of Sec. 1.2, is of particular interest because of its use in modeling more complicated surfaces through Fourier series. The surface is a square reflective surface of side length L. It differs from the previous examples in two respects. First, as a reflector, it represents an inverse aperture. Second, because we assume the material reflectance to be constant across the surface (1.0 in this case), the effect is one of phase (not amplitude) modulation on the incident light. This is because light striking a surface valley travels farther than light striking a surface peak. The grating is oriented in the xy plane at $z = 0$ with the grating lines parallel to the y axis (perpendicular to the page) as depicted in Fig. 1.1. The grating amplitude is a, the frequency is f_1, and α is the phase at $x = 0$.

$$Z(x,y) = a \sin (2\pi f_1 x + \alpha) \tag{3.32}$$

The calculation, which is given in App. B, is accomplished by finding an expression for the complex electric field at (or near) $z = 0$ and following the prescription of Eq. (3.28).

$$I(p_s) = \frac{1}{2\eta_0} \left(\frac{E_0 L^2 \cos \theta_s}{\lambda R} \right)^2 \sum_{n = -\infty}^{\infty} J_n^2(\Delta) \; \mathrm{sinc}^2 \left[L \left(\frac{x_s}{\lambda R} - n f_1 - \frac{\sin \theta_i}{\lambda} \right) \right]$$

$$\times \; \mathrm{sinc}^2 \left(\frac{L y_s}{\lambda R} \right) \tag{3.33}$$

The argument of the bessel function Δ is the peak phase retardation introduced by the grating.

$$\Delta = k_a(\cos \theta_i + \cos \theta_s) \tag{3.34}$$

The summation terms ($n = 0, 1, 2 \ldots$) represent the various orders present in the diffraction pattern. The intensities of various orders may be converted to powers and the various grating efficiencies calculated as shown in App. B. The incident power P_i is merely $(E_0 L)^2 / 2\eta_0$, where η_0 is the impedance of free space.

$$P_n/P_i = [J_n(\Delta) \cos \theta_{sn}]^2 \tag{3.35}$$

$$P_1/P_i = [\tfrac{1}{2} ka \cos \theta_{s1} (\cos \theta_i + \cos \theta_{s1})]^2 \simeq (ka)^2 \quad \text{for small angles} \tag{3.36}$$

where $J_1(\Delta) \simeq \Delta/2$ has been used.

Relative amplitudes are determined by the Bessel functions. Thus, the squared ratio of the first-to-zero-order Bessel functions (with the cosine) determines grating efficiency, which in turn can be used to find the grating amplitude a. Appendix B points out that the argument of the first sinc function, which determines order position, is nothing more than the grating equation given in Chap. 1. Order position can be used to determine the grating frequency f_1. Conversion of the measured scatter pattern to surface roughness statistics, which is discussed in Chap. 4, relies on these relationships. Notice that the surface phase term α is missing from Eq. (3.33). It appears in the corresponding expression for diffracted field strength (see App. B) and is lost in the process of taking the absolute square to obtain diffracted power. It can be recovered by interferometrically measuring the relative phase between any order and the zero order.

Equations (3.33) and (3.35) have been obtained without any apparent restrictions on the surface roughness or grating amplitude a. Actually, as pointed out by Beckmann and Spizzichino (1963, p. 178), there is a hidden restriction. The cosine term in Eq. (3.35), which comes from the obliquity factor, should actually be the cosine of the angle between the scatter direction and the surface normal, which has been assumed to be parallel to the z axis. As grating amplitude (and/or frequency) increases the amount by which the surface normal waves back and forth as a function of the x coordinate increases. Assuming that no more than a 10 percent error in the obliquity factor is permissible and that the light is normally incident on the surface, the maximum allowed grating amplitude can be shown to be

$$\frac{a_{max}}{\lambda} \leq \frac{0.025}{\pi \tan^2 \theta_s} \tag{3.37}$$

For visible wavelengths and diffraction within about 20 degrees of specular this restricts a_{max} to several hundred angstroms, which is a rough surface by optical standards. However at higher scatter angles (corresponding to steeper surface slopes) the restriction is much more severe. Although more sophisticated obliquity factors may be less sensitive, this does illustrate that one problem with the Kirchhoff theory is its ability to accurately handle high-angle diffraction. On the other hand, the Rayleigh-Rice theory, which is used in Chap. 4 to convert the diffraction pattern to surface statistics, requires that the surfaces be smoother.

The method of App. B can be applied to surfaces that are composed of more than one sinusoidal grating. For the simple case of two sinusoidal gratings (subscripts 1 and 2), oriented along the x and y axes, respectively, the diffraction intensity is given by

$$I(p_s) = \frac{1}{2\eta_0} \left(\frac{E_0 L^2 \cos \theta_s}{\lambda R}\right)^2 \sum_{n=-\infty}^{\infty} \sum_{m=-\infty}^{\infty} J_n^2(\Delta_1) J_m^2(\Delta_2)$$

$$\times \, \text{sinc}^2 \left[\left(\frac{x_s}{\lambda R} - nf_1 - \frac{\sin \theta_i}{\lambda}\right)\right] \text{sinc}^2 \left[\left(\frac{y_s}{\lambda R} - mf_2\right)\right] \tag{3.38}$$

The diffraction pattern is now composed of a grid of points in the x_s, y_s plane whose locations are determined by the arguments of the two sinc functions. These arguments are composed of small-angle versions of the two grating equations necessary to describe hemispherical diffraction. Without the small-angle approximation the pair of grating equations would have been derived exactly. Refer to Fig. 1.6 for a definition of angle ϕ_s.

$$\sin \theta_s \cos \phi_s = \sin \theta_i + f_x \lambda \tag{3.39}$$

$$\sin \theta_s \sin \phi_s = f_y \lambda \tag{3.40}$$

Observations similar to those for the one-dimensional grating apply to the calculations of grating amplitude and frequency, and to the restrictions on grating dimensions.

The true issue of boundary conditions has been avoided so far by neglecting edge effects, sticking to apertures instead of real surfaces and assuming infinite conductivity for the sinusoidal surfaces. The actual situation is more complicated. Reflectance is a function of conductivity, incident angle, and diffraction angle (and thus surface contour). It is less accurate to assume the simple modulations of the source wave used above for the more complicated case of scatter from real, arbitrarily rough surfaces.

Beckmann and Spizzichino's book (1963) gives a rather complete description of the Kirchhoff method applied to rough surfaces. They include reasonable approximations for the boundary conditions over the surface under the assumption of infinite conductivity, and treat both s- and p-polarized light. Their well-known results, which are obtained under the slightly different condition of an infinitely wide sample and are expressed as diffracted field strength over incident field strength [p. 48, Eqs. (7) and (8)], are expressed here, after notation changes, as power-grating efficiency:

$$P_n/P_i = \left[\sec \theta_i \, \frac{1 + \cos (\theta_s + \theta_{sn})}{\cos \theta_i + \cos \theta_{sn}} \right]^2 \frac{J_n{}^2(\Delta)}{J_0{}^2(\Delta)} \tag{3.41}$$

where again

$$\Delta = ka(\cos \theta_i + \cos \theta_s) \tag{3.42}$$

Except for a more complicated obliquity factor and a less complicated phase delay, the result is similar to Eq. (3.33).

The various Kirchhoff approaches allow a great deal of flexibility in dealing with different situations and are capable of predicting diffraction from known surfaces that are rougher than most optics. They cannot deal easily with exact boundary conditions on real (finite conductivity) samples and the small-angle assumptions raise questions about performance at large angles of incidence. The next section presents vector results obtained by variations on the Rayleigh approach to diffraction theory.

3.3 The Rayleigh Approach

The Kirchhoff method, outlined in the preceding section, approximates the boundary conditions present on the sample (in the aperture) and then applies some variation of the Fresnel-Kirchhoff diffraction formula [Eq. (3.20) or (3.24), etc.] to the resulting aperture field to find the field in the observation plane. In contrast, Rayleigh published a vector perturbation technique in 1895 that takes just the opposite approach. The boundary condition is left intact (almost) and the field is assumed to be composed of an infinite summation of plane waves. The solution, which takes the form of an infinite series, converges quickly only for very smooth surfaces. However, the results are applicable for most optical surfaces and can be applied to samples with finite conductivity. Beckmann and Spizzichino (1963, pps. 41, 99, 107) review the work of several authors that have published variations on the technique in radar literature (Rice, 1951; Barrick, 1970). Others followed Rayleigh's approach in following years (Maradudin, 1975; Ishimaru, 1978). Church published a series of papers based on the Rayleigh-Rice publications, that specifically addressed scatter from optical surfaces and introduced the vector perturbation technique into the optics literature (Church and Zavada, 1975; Church et al., 1977, 1979). In 1979 Elson and Bennett published a similar perturbation approach to optical scattering theory in the optical literature that proved to yield identical expressions. The technique has become known as *the Rayleigh-Rice vector perturbation theory* or sometimes just *the vector theory*. Rice succeeded in expressing the mean square value of the scattered plane-wave coefficients as a function of the surface power spectral density function. This seems quite reasonable in view of the results of the last section. Although the theoretical derivation is well beyond the scope of this book, the results have become an important scatter analysis tool and are used throughout this text.

The Rayleigh-Rice vector perturbation theory relates the scattered power density per unit incident power to the surface power spectral density function.

$$\frac{(dP/d\Omega_s)d\Omega_s}{P_i} = \left(\frac{16\pi^2}{\lambda^4}\right) \cos\theta_i \cos^2\theta_s \, Q \, S(f_x, f_y) \, d\Omega_s \qquad (3.43)$$

The quantity $(dP/d\Omega_s)d\Omega_s/P_i$ is the power scattered in the s direction through $d\Omega_s$ per unit incident power. You will notice, from Chap. 1, that, except for multiplication by the differential solid angle $d\Omega_s$ it is also the cosine-corrected BSDF. Both sides of the equation are multiplied by the differential solid angle $d\Omega_s = \sin\theta_s \, d\phi_s \, d\theta_s$ to facilitate a later integration. The quantity k^4 is sometimes referred to as *the Rayleigh blue-sky factor* because of its appearance in his explanation of molecular scattering. The cosines amount to an obliquity factor, similar to those found in the last section. The remaining quantities are used to provide a description of the sample. The dimensionless quantity Q is the reflectivity polarization factor. It expresses the action of sample material properties on the reflected light. Q is a function of the sample complex dielectric constant plus the angles of incidence and scatter, and takes on different forms depending on incident and scattered polarization states. For many cases of interest, its numerical value can be approximated by the sample reflectance. Exact relationships for Q and several approximations are discussed in Chap. 5. $S(f_x, f_y)$ is the two-sided, two-dimensional surface power spectral density function (PSD) in terms of the sample spatial frequencies f_x and f_y. As pointed out in Chap. 2 it has units of length to the fourth power. The difference of $(2\pi)^2$ between Eq. (3.43) and the corresponding equations published in early papers by Church (Church and Zavada, 1975; Church et al., 1977, 1979) is due to the notation choice of expressing the PSD frequencies as spatial cycles per unit length rather than spatial radians per unit length.

The interpretation of Eq. (3.43) is straightforward. Normalized scatter in the s direction (determined by θ_s and ϕ_s) is proportional to $S(f_x, f_y)$ evaluated at

$$f_x = \frac{\sin \theta_s \cos \phi_s - \sin \theta_i}{\lambda} \tag{3.44}$$

$$f_y = \frac{\sin \theta_s \sin \phi_s}{\lambda} \tag{3.45}$$

which are obtained from the hemispherical grating equations. The restrictions on Eq. (3.43) are those mentioned in Chap. 1. The sample must be a clean, smooth, front-surface reflector. Cleanliness and skin depth are not of concern to the theoretical nature of this chapter. The smoothness requirement restricts surface-height deviations to be much less than a wavelength and surface slope to be less than one.

$$(ka \cos \theta_i)^2 \ll 1 \tag{3.46}$$

There is no firmly established boundary; however, Fig. 3.5 illustrates the height restriction from the UV to the mid-IR using *much less than* to mean 0.01. The amplitude limit in the visible is about 100 Å. This is more restrictive than the Kirchhoff results; however, it easily meets the requirements of most mirrors. The limitation on slope is less of a problem than the height restriction for sinusoidal surfaces. High-frequency nonsinusoidal surfaces could exceed unity slopes, but this would be unusual for real surfaces. As a result, the Rayleigh-Rice relationship gives excellent results at high scatter angles.

If the surface is gratinglike (i.e., the PSD is constrained to variations in only one frequency component), Eq. (3.43) may be simplified a bit as indicated in Eq. (2.40) (Church and Zavada, 1975; Church et al., 1977, 1979). Consider $z(x,y) = z(x)$, then $S(f_x, f_y) = S(f_x) \, \delta(f_y)$ and light

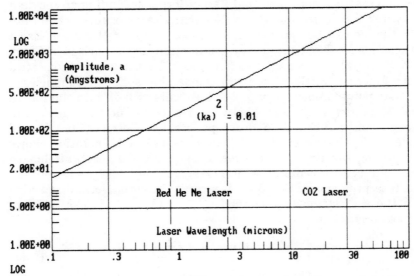

Figure 3.5 The region below the $(ka)^2 = 0.01$ limit satisfies the smooth surface requirements.

is diffracted only in directions for which $\phi_s = 0$ and, correspondingly from Eq. (3.45), $f_y = 0$. Then integrating both sides over any finite increment of ϕ_s centered about zero gives the one-dimensional analog of Eq. (3.43). Q must also be evaluated at $\phi_s = 0$.

$$\frac{[dP/d\theta_s]s\, d\theta_s}{P_i} = \frac{16\pi^2}{\lambda^3} \cos \theta_i \cos^2 \theta_s\, Q\, S(f_x)\, d\theta_s \qquad (3.47)$$

Now the light scattered into the plus-one order by a sinusoidal grating can be found by substituting in the appropriate expression for $S(f_x)$ and integrating over θ_s in the neighborhood of the diffracted spot.

$$S(f_x) = \frac{a^2}{4}\left[\delta(f_x - f_1) + \delta(f_x + f_1)\right] \qquad (3.48)$$

where $z(x) = a \sin (2\pi f_1 x + \alpha)$.

Integration over the left-hand side of Eq. (3.47) about the θ_s direction gives the grating efficiency. Integration over the right-hand side is trivial after a change of variables from θ_s to f_y. Assuming infinite conductivity, as we did in Sec. 3.2, and using the appropriate approximations for Q (to be given in Chap. 5) the grating efficiencies are

$$P_1/P_i = (ka)^2 \cos \theta_i \cos \theta_s \qquad \text{for } s\text{-polarized light} \qquad (3.49)$$

$$P_1/P_i = (ka)^2[(1 - \sin \theta_i \sin \theta_s)^2/\cos \theta_i \cos \theta_s]$$

for p-polarized light   (3.50)

For unpolarized light the two results are averaged. These are similar, but not identical to, Eqs. (3.36) and (3.41) which were derived under the same assumptions for the sinusoidal grating. Notice that under a small-angle assumption ($\theta_s \simeq \theta_i$) all three equations approach the common relationship

$$P_{s1}/P_i = (ka \cos \theta_i)^2 \tag{3.51}$$

The implication is that we can expect good agreement between the various results near specular and some divergence at higher angles.

The very general nature of Eqs. (3.43), and (3.47) and their accuracy at high angles are their main advantages over the Kirchhoff approach of the last section. Diffraction can be found for any smooth reflective sample that can be expressed as its power spectrum. And, conversely, if the BRDF of a clean, smooth reflector is known, the sample PSD can be found. This will be the topic of Chap. 4. The next section compares the various diffraction results to measured data from a sinusoidal grating.

3.4 Comparison of Scalar and Vector Results

Verification of diffraction theory is always a bit difficult because it is never clear whether modest deviations from the predicted results are due to the approximations in the theoretical model or to problems with producing a sample with a known microscopic surface. Although reasonably good sinusoidal gratings can be produced via holographic techniques, there is still a degree of uncertainty. A technique can be used that allows a comparison of the various theoretical results to measure data that are independent of grating amplitude, shape, and frequency (Stover, 1975; Schiff and Stover, 1989). Notice in all four equations [Eqs. (3.36), (3.41), (3.49), and (3.50)] that if the first-order efficiency at θ_i is normalized to itself at a fixed angle, then the grating amplitudes cancel. The various normalized efficiencies can then be plotted as a function θ_i and compared to measured efficiencies.

Figures 3.6 and 3.7 give the results of such a test. The gratings were sinusoidal surfaces with a nominal value of $a = 50$, 500, and 5000 Å and wavelengths of 6.67 μm (shown in Fig. 3.6) and 20 μm (shown in Fig. 3.7). The source light was an s-polarized HeNe laser at a wavelength of 0.633 μm. Each figure gives plots of the first-order grating efficiency as a function of incident angle, normalized by the grating efficiency at 5 degrees. The two solid curves are the results predicted

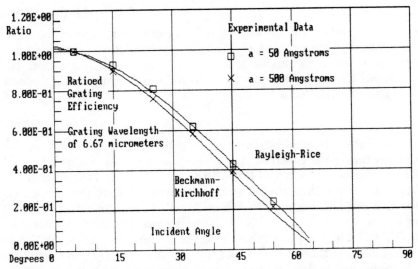

Figure 3.6 Comparison of the diffraction theories with experimental data for a 6.67 μm grating wavelength.

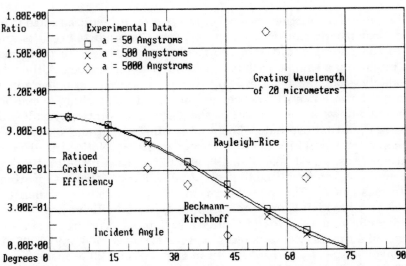

Figure 3.7 Comparison of diffraction theories with experimental data for a 20 μm grating wavelength.

by the Rayleigh-Rice approach [Eq. (3.49)] and the Beckmann-Kirchhoff approach [Eq. (3.41)]. The discrete points indicate actual measured data for the three grating amplitudes. The two theoretical results are closer for the longer spatial wavelength grating and for lower incident angles. This is expected, because as the scatter angle approaches the incident angle, the two obliquity factors become identical. The experimental fit for the 50-Å grating is actually quite good for the Rayleigh-Rice curves, regardless of incident angle, scatter angle, or grating wavelength. The 500-Å amplitude grating violates the smoothness requirement for the Rayleigh-Rice formalism, and these experimental points are found closer to the Beckmann-Kirchhoff model. The 5000-Å grating (available only at a wavelength of 20 μm) violates smoothness requirements for both gratings and neither theory predicts the wild variations in the experimental data. If shorter grating wavelengths are used, the theoretical models diverge more rapidly, and the experimental data fit very close to Rayleigh-Rice as long as the amplitudes are small (Stover, 1975). If the wavelength, or the angle of incidence, is increased the Rayleigh-Rice theory applies for the rougher gratings as well (Schiff and Stover, 1989).

3.5 Summary

The basics of diffraction theory, as it applies to the problems of light scatter, have been presented for both the Kirchhoff and the Rayleigh-Rice formulations. Experiment tells us that the perturbation approach is more accurate for smooth optical surfaces, especially at high scatter angles. This approach also has an analytical edge in terms of the versatility with which real finite conductivity samples are described. On the other hand, the mathematical derivation of the Kirchhoff approach is easier and it is more accurate for rougher surfaces. This suggests that the perturbation approach is the logical one to use for analyzing scatter data from optical reflectors and that the Kirchhoff approach should be reserved for rougher surfaces and possibly for additional mathematical analysis that may be required. In fact, that is just how they will be used in the following chapters. Chapter 4 combines the vector perturbation equations with the surface analysis of Chap. 2 to allow calculation of surface statistics from the measured BRDF.

Calculation
of Surface Statistics
from the BRDF

Chapters 2 and 3 have revealed the surface power spectral density function as the logical path to move back and forth between surface topography and surface scatter. This chapter concentrates on the application of the Rayleigh-Rice relationships to the inverse scatter problem: the calculation of reflector surface statistics from measured scatter data. Other than the scatter-measurement geometry, the details of how the scatter data is obtained is left for Chap. 6. The special cases of one-dimensional gratinglike surfaces and isotropic two-dimensional surfaces receive most of the attention. The conversion of the Rayleigh-Rice diffraction result to the Davies-Bennett TIS relationship is also reviewed. Chapters 1, 2, and 3 are used as source material.

4.1 Practical Application of the
Rayleigh-Rice Perturbation Theory

The use of scatter data as a means of specifying reflector surface quality is a powerful noncontact inspection technique. Of particular interest is the inverse scatter problem where BRDF data are used to calculate the PSD and then the various surface parameters of interest. Equation (3.43), introduced in Sec. 3.3, gives the general relationship between the surface power spectral density function (PSD) of an arbitrary, smooth, clean, front-surface reflector and the corresponding scatter pattern or BRDF. Here, the terms have been rearranged so that the BRDF is given directly in terms of measurement and sample parameters. Several points about this key result are worth mentioning.

$$\text{BRDF} = \frac{dP/d\Omega}{P_i \cos \theta_s} = \frac{16\pi^2}{\lambda^4} \cos \theta_i \cos \theta_s \, Q \, S(f_x, f_y) \qquad (4.1)$$

First, note that except for the factor $(\cos \theta_s \, Q)$, the BRDF and the surface PSD are directly proportional. Although in general, Q can change dramatically over the observation hemisphere in front of the sample, it will be learned in Chap. 5 that for the special case of an s-polarized source and plane of incidence measurements ($\phi_s = 0$ or 180 degrees) Q is given exactly by the geometric mean of the sample specular reflectances at θ_i and θ_s.

$$Q_{ss} = [R_s(\theta_i) \, R_s(\theta_s)]^{1/2} \qquad (4.2)$$

For highly reflective surfaces this means that Q is nearly equal to any measured specular reflectance. Even for the more difficult cases (Q_{sp}, Q_{ps}, out-of-plane, etc.), Q can be evaluated exactly (with effort) by using the equations presented in Chap. 5 and the value of the complex dielectric constant. Polarization effects will be discussed further in Chap. 7 and exploited in Chap. 8 to detect contaminants and subsurface defects. The important point here is that evaluation of a bandwidth-limited section of the PSD takes place directly from the BRDF without any of the integration limit problems discussed in Chap. 2.

The bandwidth-limited PSD can then be used to evaluate bandwidth-limited values for the rms roughness, the rms slope, and the average surface wavelength. Errors that arise are associated with the measurement process and the degree to which the surface meets the smooth, clean, front surface requirements and not by mathematical difficulties associated with data analysis. By the same token (as pointed out in Chap. 2), it does not make sense to attempt conversion of the calculated PSD (using a Fourier transform) to the autocovariance function, because the inherent bandwidth limits prevent integration over the required range of zero to infinity and because the same sample information is available in both functions.

Second, notice that Eq. (4.1) implies near symmetry for the BRDF. This is because the PSD is, by definition, symmetrical [see Eq. (2.29)]. For $\theta_i = 0$, the BRDF of Eq. (4.1) is exactly symmetrical. For $\theta_i > 0$, the BRDF, plotted against $\theta_s - \theta_i$ will be skewed slightly to one side relative to the specular reflection. If the BRDF is plotted against the difference of the sine of the angles ($\sin \theta_s - \sin \theta_i = \beta - \beta_0$), the symmetry is nearly exact again. It is not quite symmetrical because of the quantity $(\cos \theta_s \, Q)$. This property has been explained from several viewpoints. Church et al. (1977, 1979) points out that it is the result of conservation of momentum. Harvey (1976, 1989) takes a linear systems approach and terms it *linear shift invariance*. In view of the preceding discussions, it is clear that it can also be explained because of the near proportionality between the BRDF and the symmetrical PSD, coupled with the fact that $(\beta - \beta_0)$ is directly proportional to the spatial frequency (using the grating equation). A useful consequence of this property is that Eq. (4.1) can be used to scale the BRDF in incident angle. That is, BRDF data taken at one angle of incidence can be used to predict the sample BRDF at other angles of incidence. Figure 4.1a shows an asymmetrical BRDF ($\theta_i = 30°$) plotted against first

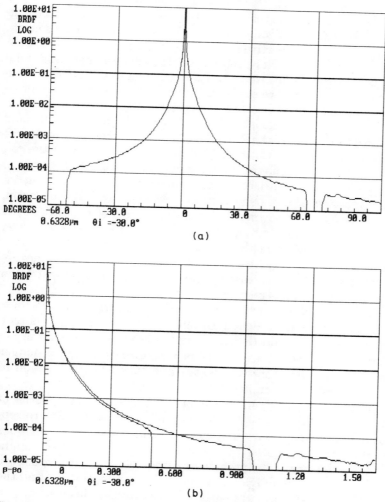

(a)

(b)

Figure 4.1   (a) The BRDF is asymmetrical when plotted against $\theta_s - \theta_i$. (b) The data in (a) exhibits near symmetry when plotted against $1\beta - \beta_0 1$. The slight deviation from symmetry is due to the factor ($\cos \theta_s\ Q$) in Eq. (4.1).

$\theta_s - \theta_i$ and then folded over (to show the symmetry) and plotted against $\beta - \beta_0$ in Fig. 4.1b.

Third, notice that Eq. (4.1) implies that scatter measurements taken at one wavelength can be used to predict scatter measurements at other wavelengths. In effect, this is done by conversion to the PSD and then reconversion to the BRDF at the new wavelength. Scaling in wavelength and angle of incidence will be discussed further in Chap. 7.

It is important to realize that these three features (calculation of the PSD, angle of incidence scaling, and wavelength scaling) are all properties of the same simple equation. They all depend on a correct conversion of the BRDF to the PSD, or in different terms, they all depend on surface topography as the only source of sample scatter. If the surface is truly smooth, clean, and front-surface reflective then all three features can be used. Conversely, if one property cannot be relied upon, then none of them can! If the sample is not *linear shift invariant*, for example, then the BRDF cannot be trusted to give correct surface statistics. Of the three, the symmetry property is the easiest to check, because its verification does not require measurement at a second wavelength or a surface-profile measurement (and the associated conversion to the PSD). The fact that symmetry (linear shift invariance) is present does not absolutely mean that surface statistics can be correctly calculated. Some transmissive samples, volume reflectors (i.e., dielectric mirrors), rough surfaces, and contaminated surfaces exhibit linear shift invariance even though they are obviously not smooth, clean, front-surface reflective. The implication is that forms of nontopographic scatter also obey the grating equation (momentum conservation) even though their intensities cannot be predicted by Eq. (4.1). Finding the scaling laws for other scatter sources, besides smooth, clean, front-surface reflectors, will dramatically increase the use of scatter measurement as an analytical tool for materials research.

Another caveat of interest is that some samples may be front-surface reflective at one wavelength but not at another. This is true for many materials at mid-IR wavelengths (see the discussion of beryllium mirrors in Sec. 7.1.2). This is not unreasonable, as longer wavelengths imply greater skin depths and increase the fraction of scatter from subsurface damage.

Most specular reflectors can be grouped into one of two categories of general interest. The first is that of isotropic surfaces which are representative of most polished reflective materials. The second includes surfaces that are dominated by one-dimensional gratinglike roughness similar in nature to the previously discussed sinusoidal gratings. These will be discussed individually under the assumption that all the requirements for calculation of surface statistics have been met.

The two-dimensional power spectrum, $S(f_x, f_y)$ is a surface above the f_x, f_y plane that can take on any symmetrical shape. This still leaves a lot of possible variation as shown in Fig. 4.2. All four PSDs pictured in Fig. 4.2 have a volcano look that peaks at low spatial frequencies (as is the case for virtually all optics). The upper rim of each volcano represents the PSD at the low-frequency limit to which each function was evaluated. A similar limit exists at f_{max}. These pictures are similar to the scatter-pattern representations shown in Fig. 1.3. In Fig. 4.2a the PSD is symmetrical in f but does not exhibit circular symmetry. The surface in Fig. 4.2b is *isotropic* in that only one sweep from f_{min} to f_{max} is necessary to characterize the entire surface because of the circular

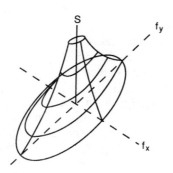

(a) A smooth non-isotropic surface.

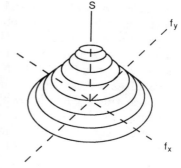

(b) An isotropic surface.

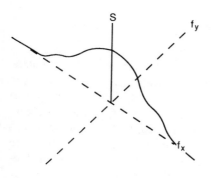

(c) A one-dimensional surface.

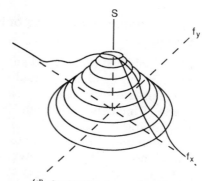

(d) Combination of a one-dimensional and isotropic surface.

Figure 4.2   PSDs of various surfaces.

symmetry. In terms of $z(x,y)$, this implies that straight-line profiles of the surface taken in any direction, from any starting point, will result in the same surface statistics if care is taken to compare over the same spatial frequency bandwidths. It does not require that $z(x,y)$ have circular symmetry and, as shown, does not restrict circular structure on the PSD. The PSD of Fig. 4.2c is of a one-dimensional surface. The last PSD in Fig. 4.2, the combination of a smooth one-dimensional surface with an isotropic surface, is similar to a precision-machined surface.

Most scatterometers take data in the incident plane in a scan that starts near, or progresses through, the specular beam, and then continues out to $\theta_s = 90$ degrees. It is rare for the full scatter hemisphere in front of the sample to be completely measured even if the instrumentation is available to do so. Single scatter measurement scans through specular correspond to radial slices in frequency space that contain the $S(0,0)$ axis and are perpendicular to the f_x,f_y plane in the pictures of Fig. 4.2. The objective is to pick the minimum number of slices (or measurements) that allow characterization of the sample. Thus, a large number of slices would be required for Fig. 4.2a and any slice will work for Fig. 4.2b. There is only one choice for Fig. 4.2c, and two slices can be used to characterize Fig. 4.2d.

Once the PSD is known, it can be manipulated according to the techniques of Chap. 2 to obtain the various roughness parameters. For example, the bandwidth-limited mean square roughness is the volume under the bandwidth-limited PSDs shown in Fig. 4.2 (i.e., the zeroth moment), the rms slope is the second moment, and so on. The next three sections discuss conversion of the scatter data to the PSD and roughness parameters.

4.2 Roughness Statistics of Isotropic Surfaces

Many polished and coated surfaces have PSDs that are nearly isotropic and can be characterized with a single sweep through the scatter hemisphere in front of the sample. The easiest choice is usually the plane of incidence slice, which corresponds to the S,f_x plane. The corresponding PSD can be found by rearranging Eq. (4.1). A factor of 10^8 has been added in Eq. (4.3) to give the PSD units of $(\text{Å } \mu m)^2$ when the wavelength is in micrometers. This allows it to be plotted over a frequency plane measured in units of inverse micrometers. The other common choice is frequency units of inverse millimeters which requires a factor of 10^{14}.

$$S(f_x, f_y) = \frac{10^8 \, \lambda^4 (\text{BRDF})}{16\pi^2 \cos \theta_i \cos \theta_s \, Q} \, \text{Å}^2 \, \mu^2 \qquad (4.3)$$

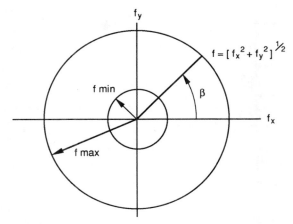

Figure 4.3 Integration of the slice around the core to obtain the effective value of the PSD, $S_{iso}(f)$.

Evaluation of Eq. (4.3) is straightforward except for the factor Q. As discussed above, analysis is simplified by making use of one of the approximations given in Chap. 5. The easiest course, to simply substitute the specular reflectance for Q, is an excellent approximation if s polarization is used and the material has a high reflectance.

To avoid the complications of a 3-D plot, the calculated slice through $S(f_x, f_y)$ can be plotted directly, or an effective value of the PSD cone, S_{iso}, can be obtained by integrating the slice around 360 degrees as shown in Fig. 4.3. The integration is trivial because $S(f_x, f_y)$ is constant for constant f.

$$S_{iso}(f) = \int_0^{2\pi} S(f_x, f_y) f d\beta = 2\pi f S(f_x, f_y) \qquad (4.4)$$

$S_{iso}(f)$ has units of length cubed. The value of f^2 at any point on the plot is the quadrature sum of the f_x and f_y components. The bandwidth-limited rms roughness σ can be found by taking the square root of the integral over f as indicated in Fig. 4.4. In most practical cases this will be evaluated as a sum of discrete values, where the distance between data points is given as Δf_i.

$$\sigma = \left[\int_{f_{min}}^{f_{max}} S_{iso}(f) \, df \right]^{1/2} \qquad (4.5)$$

$$\hat{\sigma} = \left[\sum_{i=0}^{I-1} S_{iso}(f_i) \, \Delta f_i \right]^{1/2} = \left[\sum_{i=0}^{I-1} 2\pi f_i \, S(f_i) \, \Delta f_i \right]^{1/2} \qquad (4.6)$$

$$\Delta f_i = f_i - f_{i-1} = \frac{\cos \theta_{si}}{\lambda} \Delta \theta_{si} \qquad (4.7)$$

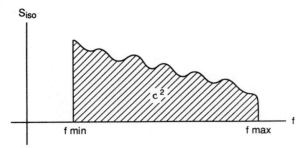

Figure 4.4 Integration of S_{iso} gives the mean square roughness.

These equations are slightly different from the estimators used in Chap. 2 to evaluate roughness because of the two-dimensional nature of the PSD, and because here it has been assumed that BRDF data result from equal increments in angle, which are not equal-frequency increments in the PSD. Using the philosophy of Chap. 2 similar relationships are obtained for the rms slope m and the average surface wavelength.

$$m = \left[\frac{1}{2} \int_{f_{min}}^{f_{max}} (2\pi f)^2 \, S_{iso}(f) \, df \right]^{1/2} \tag{4.8}$$

The factor ½ [not found in Eq. (2.32)] results from integration of the two-dimensional power spectrum.

$$\hat{m} = \left[\sum_{i=0}^{I-1} f_i^2 \, S_{iso}(f_i) \, \Delta f_i \right]^{1/2} = \left[\sum_{i=0}^{I-1} 2\pi f_i^3 \, S_i(f_i, 0) \, \Delta f_i \right]^{1/2} \tag{4.9}$$

$$\ell = 2\pi\sigma/m \tag{4.10}$$

Figure 4.5 shows a PSD plot of the molybdenum mirror of Fig. 4.1 on a log/linear scale. The integration to obtain the rms roughness is given on the linear scale to the right. Figure 4.6 shows a plot of the average surface wavelength for the same sample. In Chap. 5, this sample will be shown to exhibit scaling in both wavelength and angle of incidence.

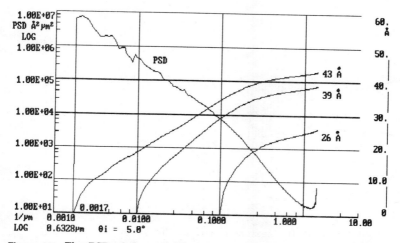

Figure 4.5 The PSD of the molybdenum mirror can be used to find the rms roughness (right-hand scale). The value of the computed roughness depends on the bandwidth of integration.

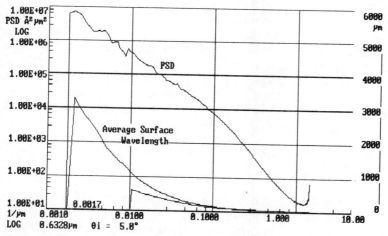

Figure 4.6 The data of Fig. 4.5 are used to calculate average surface wavelength (right-hand scale).

4.3 Roughness Statistics of One-Dimensional Surfaces

The configuration is illustrated in Fig. 4.7. The illuminating source is located in the xz plane at angle θ_i from the surface normal with the grating lines perpendicular to the incident plane. Scatter is confined

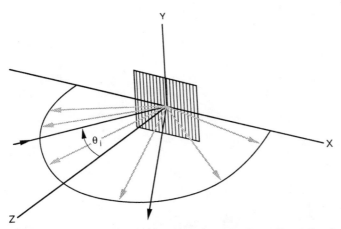

Figure 4.7 Geometry for BRDF measurement of one-dimensional surfaces.

to the incident (xz) plane. In Sec. 3.3 the one-dimensional version of the Rayleigh-Rice relationship is found by setting the two-dimensional power spectrum to its one-dimensional equivalent $[S(f_x,f_y) = S(f_x)\ \delta\ (f_y)]$ and integrating out the dependence on f_y and ϕ_s. It is appropriate to use this expression only for surfaces that are one dimensional (gratinglike). In Eq. (4.11) the terms are rearranged to solve for the one-dimensional PSD in terms of the one-dimensional BRDF (given in the brackets). The one-dimensional PSD has units of length cubed (roughness power per unit roughness frequency). As before, the factor of 10^8 has been introduced to make the units angstroms squared per unit inverse micrometer.

$$S(f_x) = \left[\frac{dP/d\theta}{P_i \cos\theta_s}\right] \frac{10^8\ \lambda^3}{16\pi^2\ Q\ \cos\theta_i \cos\theta_s}\ \text{Å}^2/\mu m \qquad (4.11)$$

The comments of the preceding section on the value of Q and the choice of polarization apply again here.

Figure 4.8 shows the BRDF for a precision-machined mirror plotted as a function of degrees from the reflected specular beam. The sample was illuminated with an HeNe laser of wavelength 0.6328 μm at an angle of incidence of 5 degrees. A number of diffraction peaks are apparent. These data are converted, via Eq. (4.11), to the PSD which is displayed in Fig. 4.9. The two plots are very similar in shape. Only the positive frequency (and positive angle) sides of the plots are shown. If both sides were plotted, the PSD would be symmetrical in frequency and the BRDF slightly asymmetrical as predicted by the grating equation (see Sec. 1.2) and the reasoning of Sec. 4.1.

Notice that the prominent peaks at frequencies of 0.45, 0.9, and 1.35

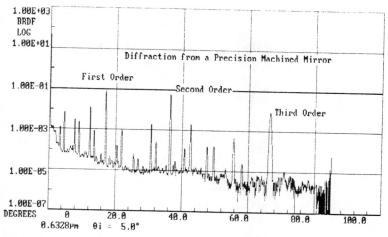

Figure 4.8 BRDF of the precision-machined mirror.

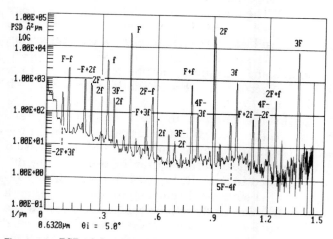

Figure 4.9 PSD of the precision-machined surface.

inverse microns in Fig. 4.9 are harmonically related. These diffraction peaks (labeled 1*F*, 2*F*, and 3*F* in the figure) are caused by the periodic tool marks left on the surface (see Sec. 2.1) by the machining process. The tool is advanced by the inverse of the fundamental frequency (2.22 μm) for each revolution of the part on the lathe spindle. The tool mark cross section is not a true cusp shape, because these peaks do not fall off as $(1/n)^4$ per the reasoning of Sec. 2.1. Instead, there is apparently more high-frequency roughness present in the cross section. This effect has been analyzed in the literature (Stover, 1976*b*) and can

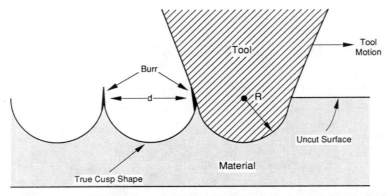

Figure 4.10 Passage of the tool (radius R) through the material with feed distance d.

be viewed interferometrically. The tip of each cusp has a small burr on its top that extends into the region that was occupied by the tool during its pass (see Fig. 4.10). The burr is folded over along the tool edge during its pass. After the tool is gone the hotter side of the burr (the tool side) cools, contracts, and the burr rises forming a long furrow on the surface (Burnham, 1976). This effect and others, which are dependent on material constants and machining parameters (feed rate, rake angle, tool radius, etc.), effect the BRDF and can be monitored with light-scatter measurements.

Several satellite peaks are grouped in pairs around the first-, second-, and third-order peaks. Two explanations can be given for the presence of diffraction peaks at these locations. The first explanation attributes their presence to the interaction (or mixing) among diffracted components explained in App. B, Eq. (B.19). It essentially reveals that when two parallel sinusoidal components are present on a surface there will be diffraction in directions that correspond to the sum and difference frequencies of the two surface waves. In this case, as described above, the prominent tool marks are not sinusoidal, so there are several harmonically related surface waves. When the harmonics of one fundamental are summed and differenced with each other the resulting frequencies are again multiples of the fundamental so no new frequencies are expected. A direct analog of this behavior will be familiar to electrical engineers with a background in communications theory. Careful study of Fig. 4.9 reveals that a second harmonically related series (labeled $1f$, $2f$, $3f$) starts at 0.33 inverse microns. When this series of surface waves mixes with the series starting at 0.45 inverse microns, a variety of new peaks at various sum and difference frequencies are created in the diffraction pattern. All of the major peaks in Fig. 4.9 can be identified as a combination of these two series as indicated. So this explanation does predict the correct location of these peaks. Equation (B.19) also predicts the relative intensities of the various peaks. Using the notation of Eq. (B.19), each peak intensity is given by $J_n^2(\Delta_F)J_n^2(\Delta_f)$. The two Bessel series in n and m share the same zero order (that is, $J_0^2(\Delta_F) = J_0^2(\Delta_f)$. Setting the cosines in the obliquity factor and the Bessel function arguments equal to unity (i.e., assuming small-angle scatter) gives the first-order diffracted intensities at 0.33 and 0.45 inverse micrometer values of $J_0^2(\Delta_F)J_1^2(\Delta_f)$ and $J_0^2(\Delta_f)J_1^2(\Delta_F)$, respectively. The sum and difference peaks (labeled $F - f$ and $F + f$) have calculated intensities given by $J_1^2(\Delta_f)J_1^2(\Delta_F)$. Approximate values for $J_1^2(\Delta_F)$ and $J_1^2(\Delta_f)$ may be found to be 3×10^{-3} and 5×10^{-4} using Fig. 4.8 which includes the BRDF at zero. The sum and difference intensities (at $F + f$ and $F - f$) should be lower than those at F and f by these multipliers if the calculation of App. B is to be used to explain their existence. Examination of Fig. 4.8 quickly confirms that this is not the case. The sum and difference terms are larger than expected by about three orders of magnitude. The difference is too large to be caused by our cavalier treatment of the cosines. Hence, only a small fraction of these satellite peaks is due to the nonlinear mixing of App. B.

The second possible source of these peaks in the diffraction pattern is their actual appearance as sinusoidal components on the surface. This means that the machine tool is somehow producing them. In effect, the nonlinear mixing between the vibration f and the feed F takes place in the machine tool and the resulting motion of the tool on the part faithfully reproduces these components. There was apparently a machine vibration present causing relative motion between the part and the tool, that created the 0.33 inverse micrometer fundamental. Its time frequency, in hertz, can be found from the spindle speed used when cutting the part. This can be used as a clue to finding the vibration source and eliminating it. This particular BRDF is an indication that the machine tool itself has a serious problem and represents a source of useful production feedback that goes beyond roughness characterization of the surface.

The rms roughness, slope, and average wavelength are again found by the integration techniques of Chap. 2 and summed over I samples.

$$\sigma = \left[2 \int_{f_{min}}^{f_{max}} S(f_x)\, df \right]^{1/2} \tag{4.12}$$

$$\hat{\sigma} = \left[2 \sum_{i=0}^{I-1} S(f_i)\, \Delta f_i \right]^{1/2} \tag{4.13}$$

$$m = \left[2 \int_{f_{\min}}^{f_{\max}} (2\pi f)^2 \, S(f_x) \, df \right]^{1/2} \tag{4.14}$$

$$\hat{m} = \left[2 \sum_{i=0}^{I=1} (2\pi f_i)^2 \, S(f_i) \, \Delta f_i \right]^{1/2} \tag{4.15}$$

$$\ell = 2\pi\sigma/m \tag{4.16}$$

In Fig. 4.11 the surface rms roughness of the same part is shown superimposed on the PSD. The integral starts on the left near $f = 0$ and progresses toward the roughness scale on the right-hand side of the plot. The abrupt contributions of the various diffraction peaks and the bandwidth-limited nature of the rms roughness are evident. The peak amplitudes of individual sinusoidal components can be evaluated from the corresponding rms contributions. For example, the fundamental at 0.45 inverse micrometers increases the rms integral from 12.5 to 22.5 Å. Remembering that these contributions add linearly to the mean square roughness, gives

$$\sigma(0.45) = (22.5^2 - 12.5^2)^{1/2} = 18.7 \text{ Å} \tag{4.17}$$

which converts to a peak sinusoidal amplitude of 26.5 Å. The average surface wavelength, shown in Fig. 4.12, is also strongly affected by the presence of the prominent periodic components.

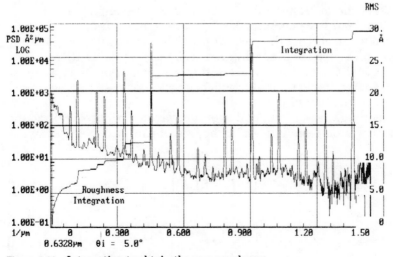

Figure 4.11 Integration to obtain the rms roughness.

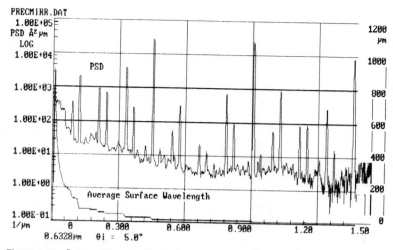

Figure 4.12 Integration to obtain the average surface wavelength.

Of course the precision-machined surface illustrated above is not truly one dimensional. All real surfaces scatter some light throughout the full-scatter hemisphere, which implies that there are roughness components that do not run parallel to the tool marks. These components are analyzed by rotating the sample 90 degrees about its surface normal and taking a BRDF scan perpendicular to the plane of the prominent diffraction peaks. These data are analyzed using the isotropic assumption of the last section. The resulting PSD is shown in Fig. 4.13 with the corresponding integration for the rms roughness. As expected, most of the surface roughness was associated with the one-dimensional components. In order to get the total rms roughness, the values must be added in quadrature.

$$\sigma_{total} = (\sigma_{1D}^2 + \sigma_{iso}^2)^{1/2} = (30^2 + 4^2)^{1/2} = 30.2 \text{ Å} \qquad (4.18)$$

It is important to make certain that the two values of σ (one dimensional and isotropic) correspond to the same band of spatial frequencies before they are added.

4.4 Roughness Statistics for the General Case

Occasionally sample scatter needs to be measured over the full observation hemisphere in front of the sample. The BRDF at any point can be used to find the corresponding values of the PSD, via Eq. (4.3), and the surface statistics can again be found using the results of Chap. 2.

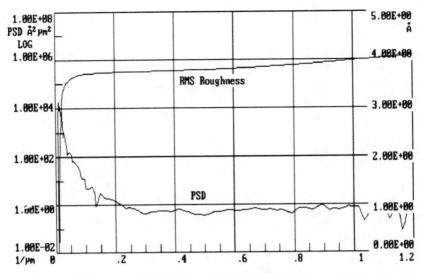

Figure 4.13 The isotropic PSD of the precision-machined mirror of Figs. 4.8, 4.9, and 4.11.

$$\sigma = \left[\int\limits_{f_{\min}}^{f_{\max}} \int S(f_x, f_y) \, df_x \, df_y \right]^{1/2} \tag{4.19}$$

$$df_x \, df_y = \frac{\cos \theta_s \, d\Omega_s}{\lambda_2} = \frac{\cos \theta_s \sin \theta_s \, d\phi_s ds\theta}{\lambda^2} \tag{4.20}$$

$$\hat{\sigma} = \left[\sum_{n=1}^{N} \sum_{m=1}^{M} S(f_x, f_y) \left(\Delta f_{xn} \, \Delta f_{ym} \right) \right]^{1/2} \tag{4.21}$$

$$\Delta f_{xn} \, \Delta f_{ym} = \frac{\cos \theta_s \sin \theta_s \, \Delta \phi_s \, \Delta_s \theta}{\lambda^2} \tag{4.22}$$

The frequency "band" of interest is now an area on the f_x, f_y plane. The limits of integration are determined by the shape of the area. It is often convenient to use polar coordinates. As pointed out in Chap. 2, the concept of surface slope on a two-dimensional surface is of limited value without a well-defined direction unless the surface is isotropic.

In practice the situation is not always this simple because the scatterometer receiver may be swept over the hemisphere using rotations about axes other than the θ_s and ϕ_s axes (see Fig. 6.13). In this case, an instrument's specific transformation must be added to the above equations to convert from the instrument coordinate system to the analysis coordinate system.

4.5 The K-Correlation Surface Power Spectrum Models

Many surface power spectrums have shapes that are close fits to an algebraic form. As an example notice that the PSD for the molybdenum mirror of Fig. 4.5 is close to a straight line on a log-log plot. The K-correlation model is commonly used for this purpose (Church et al., 1989) and can be expressed for both one- and two-dimensional profiles in terms of the parameters A, B, and C.

$$S_1(f_x) = \frac{A}{[1 + (Bf_x)^2]^{c/2}} \qquad (4.23)$$

$$S_2(f) = \frac{A'}{[1 + (Bf)^2]^{(c+1)/2}} \qquad (4.24)$$

where

$$A' = \frac{\Gamma[(c+1)/2]}{2\sqrt{\pi}\Gamma(c/2)} AB \qquad (4.25)$$

Curves that take this shape look like low-pass filters. The value of A is determined by low-frequency behavior (i.e., small Bf). The parameter B is related to the correlation length. For the special case $C = 2, B$ equals $2\pi\ell_c$, where ℓ_c is the e^{-1} definition of the correlation length. In effect, B determines the frequency location of the break point in the low-pass response, that separates the low- and high-frequency regions. The value C determines the rate of falloff (or slope) of the power spectrum at high frequencies. Two special cases have found prominence in the literature.

4.5.1 The Lorentzian power spectrum

If $C = 2$ the K-correlation form is called a Lorentzian. Workers at the University of Arizona Optical Science Center have used this form to reduce BRDF data to surface statistics (Wolfe and Wang, 1982; Wang, 1983). The relationships for one- and two-dimensional surfaces are given below:

$$S(f) = \frac{2\sigma^2 \ell_c}{1 + (2\pi f \ell_c)^2} \tag{4.26}$$

$$S(f_x, f_y) = \frac{2\pi\sigma^2 \ell_c^2}{[1 + (2\pi f \ell_c)^2]^{3/2}} \tag{4.27}$$

These curves can be conveniently divided into the two sections $2\pi f \ll 1$ and $2\pi f \gg 1$. In the low-frequency section, the curve is essentially constant. In the high-frequency section the curve has a constant negative slope when $\log(S)$ is plotted against $\log(f)$. This allows the power spectrum to be fit with two straight-line asymptotes whose constant value and constant slope (respectively) can in principle be easily evaluated. The technique, which is outlined in Fig. 4.14, merely consists of finding the breakpoint, or knee (S', f'), at which the two asymptotes meet and evaluating the two unknowns in terms of the equations of the asymptotes. The technique is interesting in that many samples do exhibit constant slopes on log-log plots. Unfortunately the Lorentzian equations above call for slopes of -2 and -3 for the one- and two-dimensional cases, respectively, and this is often not the situation. Even more troublesome, from a practical point of view, is that the breakpoint can often not be found. It is at a small enough frequency (corresponding to near-specular scatter) that the measurement technique does not always reveal it. These limitations severely restrict the usefulness of this technique for evaluation of surface statistics, especially in view of the ease with which results can be obtained with the

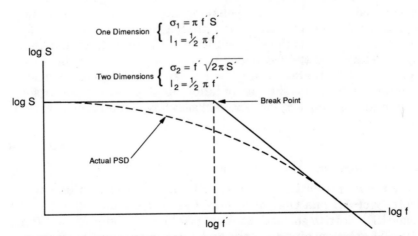

Figure 4.14 Solution for the rms roughness and autocovariance length for the special case of a Lorentzian power spectrum.

methods of Secs. 4.2 to 4.4. However, the idea of analyzing BSDF data that exhibit constant log-log slopes will be pursued further in the next section.

4.5.2 Fractal surfaces

Some optically finished surfaces exhibit the curious property that their measured, or calculated, surface power spectrums very nearly follow an inverse power law with no apparent breakpoint (that is, $Bf \gg 1$). This was pointed out by Church et al. (1979) and the measurement implications of such behavior have since been considered (Church, 1980, 1988; Church et al., 1989). For one-dimensional surfaces the power spectrum is expressed as

$$S(f_x) = \frac{K_n}{f_x^n} \tag{4.28}$$

where $1 < n < 3$ and K_n is a constant with the units of $(\text{length})^{3-n}$. Here the value $n = C$ has been substituted and K_n is a combination of A, B, and C in accordance with some literature notation. This means that a log-log plot of the power spectrum is a straight line with slope $(-n)$ and value (K_n) at $f_x = 1$. If the surface is isotropic the two-dimensional power spectrum takes the following form:

$$S(f) = \frac{\Gamma[(n + 1)/2]\, K_n}{2\Gamma(1/2)\, \Gamma(n/2)\, f^{n+1}} \tag{4.29}$$

These are called fractal surfaces and they have some interesting properties. The special cases $n = 1$, 2, and 3 are called the extreme fractal, the Brownian fractal, and the marginal fractal, respectively. Notice that there is a problem at $f = 0$—the power spectrum cannot take on an infinite value. In reality, it must roll off at some low frequency as predicted by the full three-parameter K-correlation model. This makes the form of these PSDs very similar to the high-frequency section of Lorentzian shapes discussed earlier, except that the slope is no longer restricted to -2 or -3 for the one- and two-dimensional cases respectively.

A true fractal profile (with breakpoint) will have a correlation length approximately equal to the length of the profile under consideration. And the rms roughness calculated from such a PSD will be strongly dominated by the low-frequency contributions. Thus, the calculated roughness parameters are strongly bandwidth (or measurement) dependent for fractals, just as they are for nonfractals. However, fractals have the unique quality that their power spectrums can be characterized by only two quantities—n and K_n. In other words, in some ways, it would make more sense to report the fractal constants instead of a height parameter (rms roughness), transverse parameter (average wavelength or correlation length), and the associated bandwidths.

4.6 The TIS Derivation from the Rayleigh-Rice Perturbation Theory

When Davies (1954) derived the relationship between smooth-surface total integrated scatter (TIS) and the rms roughness (Sec. 1.6) he assumed that most of the scatter was close to specular (cos θ_s = 1 for θ_i = 0) and that the surface had a Gaussian height distribution. Both assumptions simplify the mathematics. The Gaussian assumption was used as a convenient scapegoat for several years to explain differences in measured rms roughness observed when the same sample was measured by different measurement techniques. The Rayleigh-Rice equations, which relate scatter to the surface power spectrum independent of the form of the height distribution, suggest that perhaps the Gaussian assumption was not necessary. Church pointed this out in 1977 and his approach is outlined here.

The TIS is first expressed in terms of the BRDF by integrating from the small entrance/exit hole out to the waist of the observation hemisphere (see Fig. 1.7). The polarization constant is approximated by the surface reflectance R and then Eq. (4.1) is used to express the BRDF in terms of the surface power spectrum. An exchange of variables, from θ_s, ϕ_s to f_x, f_y is made to allow integration over the power spectrum. The small scatter-angle assumption is then made, which removes the cosines, and the result is the familiar expression for TIS in terms of the surface rms roughness.

$$\text{TIS} \equiv \frac{\text{scattered power}}{\text{specularly reflected power}} = \int_0^{2\pi} \int_{\theta_{min}}^{\pi/2} \frac{(dP/d\Omega)_s}{RP_i} \, d\Omega_s \quad (4.30)$$

$$\text{TIS} = \left(\frac{4\pi}{\lambda}\right)^2 \int\int_{f_{min}}^{f_{max}} \cos\theta_i \cos\theta_s \, S(f_x, f_y) \, df_x df_y \quad (4.31)$$

where

$$df_x df_y = \frac{d\Omega_s \cos\theta_s}{\lambda^2} \quad (4.32)$$

$$\text{TIS} = \left(\frac{4\pi}{\lambda}\right)^2 \int\int_{f_{min}}^{f_{max}} \cos\theta_i \cos\theta_s \, S(f_x, f_y) \, df_x df_y \approx \left(\frac{4\pi\sigma \cos\theta_i}{\lambda}\right)^2 \quad (4.33)$$

The result is important from several aspects. In addition to removing the Gaussian restriction, it is another point of common ground between the Rayleigh-Rice and Kirchhoff approaches to diffraction theory. The fact that TIS is strictly a scalar result is brought home by the use of the specular reflectance in place of the polarization constant. Scatter amplitude from a normally illuminated isotropic sample will not be constant in ϕ_s at fixed θ_s if the source is plane polarized. The variations are due to the differences in s and p polarization that are evident in the expressions for Q. These differences can amount to more than those imposed by the small-angle assumption and the problems with changing angle of incidence on the TIS detector (Stover and Hourmand, 1984a).

A simple demonstration of the effectiveness of TIS measurements on non-Gaussian surfaces can be made by returning to the sinusoidal grating (which is obviously non-Gaussian). The scatter signal will consist of two first-order diffraction spots as predicted by the Rayleigh-Rice result [Eq. (3.49)] for s polarization.

$$P_s/P_0 = 2[(ka)^2 \cos \theta_i \cos \theta_s]|_{\theta_s \approx \theta_i \approx 0} \approx 2 \left(\frac{4\pi a}{\lambda}\right)^2 = \left(\frac{4\pi\sigma}{\lambda}\right)^2 \qquad (4.34)$$

Assuming unity reflectance, normal incidence, and small-angle scatter, and substituting the sinusoidal rms roughness $\sigma = a/\sqrt{2}$ gives the Davies' TIS result. Even the assumption of s polarization is not critical since both polarization results are identical for small-angle scatter [Eq. (3.51)].

4.7 Summary

When the smooth, clean, front-surface reflective conditions are met, a bandwidth-limited section of the surface power spectral density function (PSD) is nearly proportional to the angle-limited BRDF. Given one of the two functions, it is then possible to find the other. If the PSD is found from the BRDF, it is possible to compute the surface rms roughness, rms slope, and average surface wavelength as indicated in Chap. 2. The specific equations of interest have been presented for the special cases of the isotropic and one-dimensional surfaces which can be analyzed using plane-of-incidence scatter data. Equations for the more general case, requiring scatter measurement over the full hemisphere in front of the sample have also been given. Power spectrums that fit (or nearly fit) K-correlation expressions offer the advantage of characterizing the surface with quantities that are not measurement (bandwidth) dependent. The Rayleigh-Rice relationship has been integrated to obtain the familiar relationship between TIS and rms roughness without assuming that the surface has a Gaussian height distribution. Finally, the philosophical stage has been set to introduce wavelength and angle-of-incidence scaling—which will be done in Chap. 7.

Polarization
of Scattered Light

Some of you are probably old enough (like me) to remember when polaroid sunglasses first became available. They were a sensation. Glare light is reduced by more than the background light with the result that you actually have better vision—not just less light in your eyes. They work best in situations where the sun is more or less in front of you and the combination of reflected light and near-specular scatter causes a bright glare that dominates your field of view. Because reflectance is a function of polarization, the glare light (which has been reflected once) often has a horizontal polarization component that is much stronger than the vertical component. Normal background light (which has been reflected many times at many angles) is more evenly divided between the two polarizations. Polaroid sunglasses simply discriminate against the horizontal component to reduce the fraction of glare light. And if you look at the blue sky through your sunglasses, you will find that rotating the glasses 90 degrees causes the sky to look dimmer (i.e., the sky has a strong vertical component). This effect is most strongly pronounced if you look in a direction perpendicular to the sun's rays. These two examples clearly demonstrate that scattering is polarization sensitive.

When light is scattered, its polarization, as well as its amplitude and direction, are changed. The changes depend on the sample shape and material, as well as the polarization, amplitude, and direction of the incident beam. All three quantities must be considered to examine the effect of a sample on the reflected and transmitted light. A complete description of the polarization characteristics of an EM wave, before or after sample interaction, can be accomplished by straightforward measurements. Comparison of polarization amplitude and direction, before and after interaction with the sample, allows information about the sample material to be obtained. The trick is to relate the before and after changes to useful, or needed, sample characteristics.

Chapters 3 and 4 discussed the relationship between smooth reflector topography and the resulting changes in the reflected light. Polarization changes were presented as contained in the factor Q, which for a convenient choice of incident polarization and incident plane measurements is nearly constant. For that particular application, the correct choice of polarization made surface characterization straightforward. There are other scatter-measurement applications, besides roughness characterization, that benefit from analysis of the polarization state of both incident and reflected light. Different forms of characterizing polarization are often used in these situations. This chapter reviews polarization concepts and characterization requirements in a general sense using various scattering vectors and matrices. The factor Q is then completely defined and discussed in terms of its general description. The use of wave-polarization vectors and matrices to represent sample/component effects is reviewed. These representations are shown to be more tedious, and less quantitative, in terms of the sample's physical characteristics, but they can be extremely useful for locating and empirically grading sample defects and changes in sample properties. In other words, polarization effects present a noncontact technique to monitor quality in process control systems and in final inspection.

5.1 A Review of Polarization Concepts

The assumption behind most of this text has been that the reader does not need a tutorial on the basic principals of optics. In this section that assumption is dropped long enough to briefly review required polarization concepts. This is done to save a fairly large fraction of the readers a dash for their reference texts, and to avoid confusion due to the diversity in the way various terms have been defined in the literature. Some readers will want to skim this section just long enough to pick up the nomenclature. For those that want a more complete review of polarization, a number of texts are available (Shurcliff, 1962; Jenkins and White, 1976; plus many others). Appendix A reviews necessary wave propagation concepts.

It is the electric component, in the transverse electromagnetic wave description that is responsible for most observed EM wave/material interactions. And, it is the direction of this vector that is used (in this text) to define the direction of polarization. In general, the polarization of monochromatic coherent light is elliptical. As shown in

Fig. 5.1, all of the elliptical configurations can be expressed as the summation of two orthogonal linearly polarized waves. At zero phase difference ($\delta = 0$) the two waves sum to a linearly polarized beam. If the phase difference is increased, the resultant polarization moves from linear, to elliptical, to circular, back to elliptical and at a phase difference of π the resultant wave is plane polarized again, but rotated by 90 degrees from the original zero-phase condition. Increasing the

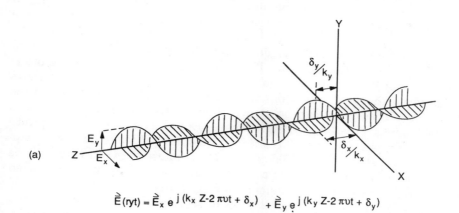

$$\vec{E}\,(ryt) = \vec{E}_x\, e^{\,j\,(k_x\, Z - 2\pi \upsilon t + \delta_x)} + \vec{E}_y\, e^{\,j\,(k_y\, Z - 2\pi \upsilon t + \delta_y)}$$

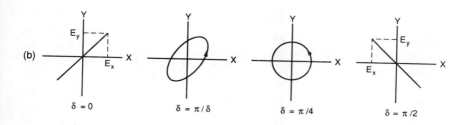

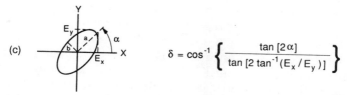

$$\delta = \cos^{-1}\left\{\frac{\tan\,[2\alpha]}{\tan\,[2\,\tan^{-1}(E_x/E_y)]}\right\}$$

Figure 5.1 (a) Two orthogonal plane waves combine to form an elliptically polarized wave. (b) Viewed from the $+Z$ axis the resultant vector sweeps out an elliptical form in x,y that depends on $\delta = \delta_y - \delta_x$, E_x, and E_y. (c) The value of δ may be obtained from E_x, E_y, and α as shown.

phase difference by another π (a total of 2π) brings the wave back to its original plane-polarized state. Phase differences may be introduced between wave components through the use of retardation plates (quarter wave, half wave, etc.) made of birefringent materials. The direction around the ellipse that is traced out by the resultant electric vector, is determined by which of the two plane waves leads in phase. If the vector rotates in the clockwise direction when propagating toward the observer, it is right-handed polarization. Counterclockwise rotation is known as left-handed. Using the phasor notation of Chap. 3 and App. A, the figure shows that the amplitudes of the two plane waves (E_x, E_y) and the phase difference between the two $(\delta = \delta_x - \delta_y)$ are the three quantities needed to completely characterize the general elliptical polarization state. Thus for linear optics, all waves of interest, regardless of the polarization state, can be represented by the sum of two orthogonal, out-of-phase, linearly polarized waves. Without too much difficulty, it can be shown that arbitrarily phased right- and left-handed circularly polarized waves can also be used as a composite pair to represent the polarization state of an EM wave. Figure 5.1 applies to a strictly polarized source. If the phase between E_x and E_y is not well defined, as in the case of a quasimonochromatic wave, then the light is said to be unpolarized or partially polarized.

The polarization characteristics of scattered light are more easily understood in terms of the concepts developed for specular light. Figure 5.2 shows the geometry for light P_i incident on a sample medium of index n with a boundary surface at the x,y plane. The boundary reflects P_r, transmits P_t, and scatters P_s. Each ray has a plane of propagation defined by the ray direction (k vector) and the surface normal (z). The electric field vectors (E_i, E_r, E_t, E_s) are composed of an s component, perpendicular to the plane of propagation, and a p component that is in the plane of propagation. Notice that the s components are all parallel to the sample face (the xy plane). The other common notation for these two polarizations is $\perp$ for s and $\parallel$ for p. These are more functional than x and y because polarization needs to be defined in terms of the beam-propagation plane, which depends on both beam direction and sample orientation. The arrangement in Fig. 5.2 is essentially the same geometry used to define the BSDF in Chap. 1. There are some serious issues regarding sample orientation within this geometry but they will be put off until Chap. 6.

When the situation of Fig. 5.2 is analyzed by applying the electromagnetic boundary conditions, two familiar results are derived. The first is Snell's law, which is a result of the condition that the phase variations of the incident, reflected, and transmitted waves at the interface be identical. This relationship may be used to find the angle from normal of the transmitted light.

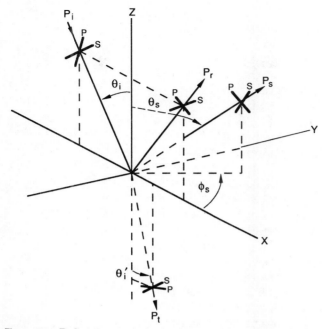

Figure 5.2 Definition of s and p polarization in terms of the incident plane and scatter plane.

$$n \sin \theta_t = \sin \theta_i \tag{5.1}$$

The angles found through Snell's law are independent of polarization if the index of refraction is independent of polarization. The second result, known as the Fresnel reflection equations, relates the field strengths on either side of the boundary. The subscripts i, r, and t are used to indicate incident, reflected, and transmitted fields and powers, respectively. The s and p subscripts refer to the polarization state. These equations may be used to establish the relative values of P_t, P_r, and P_i and their dependence on the angle of incidence, the index of refraction and the polarization.

$$\frac{E_{rs}}{E_{is}} = -\frac{\sin(\theta_i - \theta_t)}{\sin(\theta_i + \theta_t)} = \frac{\cos\theta_i - \sqrt{n^2 - \sin^2\theta_i}}{\cos\theta_i + \sqrt{n^2 - \sin^2\theta_i}} \tag{5.2}$$

$$\frac{E_{rp}}{E_{ip}} = \frac{\tan(\theta_i - \theta_t)}{\tan(\theta_i + \theta_t)} = \frac{n^2\cos\theta_i - \sqrt{n^2 - \sin^2\theta_i}}{n^2\cos\theta_i + \sqrt{n^2 - \sin^2\theta_i}} \tag{5.3}$$

$$\frac{E_{ts}}{E_{is}} = \frac{2\sin\theta_t\cos\theta_i}{\sin(\theta_i + \theta_t)} = \frac{2\cos\theta_i}{n^2\cos\theta_i + \sqrt{n^2 - \sin^2\theta_i}} \tag{5.4}$$

$$\frac{E_{tp}}{E_{ip}} = \frac{2 \sin \theta_t \cos \theta_i}{\sin (\theta_i + \theta_t) \cos (\theta_i - \theta_t)} = \frac{2n \cos \theta_i}{n^2 \cos \theta_i + \sqrt{n^2 - \sin^2 \theta_i}} \quad (5.5)$$

The value n is the refractive index of the medium beyond the interface divided by the index of the incident beam medium. For light in space, incident on a dielectric, n becomes the refractive index of the dielectric. The reflection equations will be considered first. The minus sign in front of the ratio in Eq. (5.2) implies a 180-degree phase shift of the reflected s-polarized component. The dependence on the index of refraction can be made apparent by expanding the different functions into products, and using Snell's law to convert from transmitted angles to incident angles. The squares of the first two relationships for the reflected power are plotted in Fig. 5.3 for the case of $n = 1.5$. The dip to zero, at 56.3 degrees for P polarization, is called Brewster's angle (or the polarization angle). Brewster's law gives the polarization angle as $\tan \theta_i = n$. Equation (5.3) predicts its existence at $\theta_i + \theta_t = 90°$ or when the reflected and transmitted polarizations are 90 degrees apart. This makes sense physically because the reflected ray would have to travel in the direction of the transmitted p-polarized electric field vector, which is caused by vibrations of the induced material dipoles. Because light is a transverse wave, propagation in the direction of the induced electric field cannot take place.

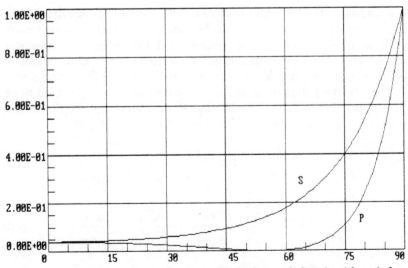

S and P Reflectances

Figure 5.3 Reflectance of s- and p- polarized light from a dielectric with an index of 1.5.

Total internal reflection may also be predicted from these equations. In this case the wave is crossing an interface that goes from high index to low index. The effect on the equations is that n, which is really the ratio of indices, becomes less than one. For both s and p polarization the reflectance is equal to one when the square root term goes to zero. This corresponds to the transmitted beam being refracted along the interface surface.

Thus, in general, if the source is unpolarized, a lot more s-polarized light will be found after reflection. It is this light that is filtered out by polaroid sun glasses. The reflectance at both polarizations is unity at grazing angles. Finally, notice that Fig. 5.3 is a plot of the ratioed intensities (field strength squared). Because the incident and reflected beams are traveling in the same medium (at the same speed) and because they have the same cross-sectional area, the intensity ratio is equal to the power ratio. This is not the case for the transmitted beam traveling in index n. Its cross-sectional area is increased by the ratio of the propagation direction cosines after refraction, and its time-average power flow is proportional to n [see Eq. (A.23) in App. A]. Thus, assuming no absorption, the power conservation equation for either polarization becomes

$$\left(\frac{E_r}{E_i}\right)^2 + n \left(\frac{E_t}{E_i}\right)^2 \frac{\cos \theta_t}{\cos \theta_i} = R + T = 1 \qquad (5.6)$$

Because more s-polarized light is reflected, one would expect more p-polarized to be transmitted. This is easily verified by noticing that Eqs. (5.4) and (5.5) are identical except for the $\cos (\theta_i - \theta_t)$ factor. Finally, the Fresnel equation can be evaluated at zero angle of incidence. If the sample is isotropic there is no distinction between s- and p-polarized specular light—there are no asymmetries. Converting from intensities to powers, as indicated above, gives

$$R = \left(\frac{n - 1}{n + 1}\right)^2 \quad \text{and} \quad T = \frac{1}{n}\left(\frac{2n}{n + 1}\right)^2 \qquad (5.7)$$

which evaluate to 0.04 and 0.96, respectively, for $n = 1.5$—a value which is representative of many dielectrics at visible wavelengths.

Several preliminary conclusions can be drawn about the polarization dependence of the scatter pattern associated with the reflection and refraction of light from an isotropic dielectric plane. Based on the relative values of reflectance and transmittance, scatter in the forward direction (transmission) is likely to exceed scatter in the backward (reflective) direction. Thus more signal can be expected in the forward direction when scatter is used to locate defects in transparent materials. For an unpolarized input, there is likely to be a lot more s-polarized light in the reflected scatter than p-polarized light. The reverse is true of the transmitted scatter pattern. These statements are true simply because a larger fraction of the incident p-polarized light is transmitted. Although at zero angle of incidence there is a polarization symmetry for the specular beams, this is not true for the scattered beams. If a plane-polarized beam is normally incident on the sample, there is an asymmetry over the scattering sphere relative to the direction of polarization. And further, based on the reasoning used to explain the Brewster angle, there should be scatter directions in which there is only one polarization allowed.

Figure 5.4 compares reflective and transmissive scatter from a glass window. The scatter was observed to be strong at the first sample surface, weak at the second sample surface, and very weak in the bulk. The incident light was a circularly polarized HeNe laser at a wavelength of 0.633 μm and an incident angle of 30 degrees. Both s and p scatter were measured on each side of the window, as indicated in Fig. 5.4a, by placing a polarizing filter (analyzer) in front of the detector. Using the transmitted (or reflected) specular beam as 0 degrees the data in Fig. 5.4b was obtained. As predicted the reflective s-polarized scatter was stronger than the p-polarized scatter. The second surface reflection is also apparent in the reflective scans just to the right of the main specular peak. The two transmissive scans are almost identical in intensity. This is because the reflective component removes a relatively small fraction of the light from the transmitted beam [R_s = 0.058, R_p = 0.025 using Eqs. (5.2) and (5.3)] so the ration of s to p light in the transmitted beam is still nearly equal. The transmissive scatter is modulated by another effect described by the Fresnel equations. The fast drop in transmissive scatter between 10 and 20 degrees is caused by total internal reflection of front-surface scatter at the second surface. This brings up another point. Because the reflection from the high/low index direction is greater than from the low/high index direction, windows should be oriented with the high-scatter surface away from the hemisphere of concern.

The situation for reflection from metals is more complicated because, unlike dielectrics, a phase difference is introduced between the s- and p-reflected components. The degree of elliptical polarization introduced into the reflected beam is a function of the angle of incidence, as well as the wavelength-dependent metallic optical constants. These effects are explained through the use of a complex refractive index (or complex dielectric constant), which is given in terms of the optical constants and depends on the material conductivity (see App. A). Metallic reflectance at normal incidence is generally higher than for a dielectric. There is a nonzero minimum in the p-polarized reflectance, similar to the Brewster angle minimum for dielectrics, called the prin-

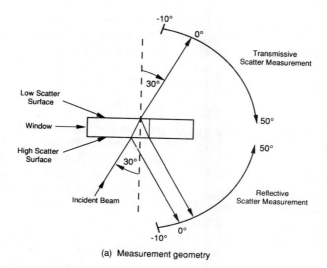

(a) Measurement geometry

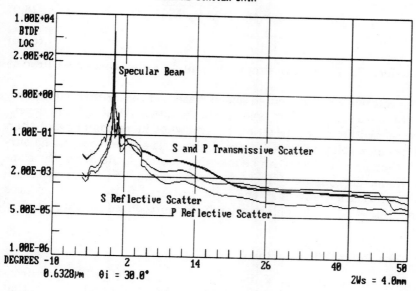

(b) Measured scatter

Figure 5.4   Scatter measured from a glass window.

cipal angle of incidence, $\theta_{i'}$. By choosing an appropriate configuration, the complex index can be evaluated after appropriate measurement of the polarization state of both the incident and reflected specular beams. This process, which is called ellipsometry, is based on Eqs. (5.2) and (5.3), where the complex index is given as

$$\hat{n} = n + jnK = n + jK_0 \qquad (5.8)$$

The real index n and the absorption index K (or the absorption coefficient K_0) are known as the optical constants. Taking the absolute squares of Eqs. (5.2) and (5.3) gives the s and p reflectances as a function of incident angle. By taking two reflectance measurements these relationships can be used to produce two equations which can be solved for the two unknowns. Many measurement combinations are possible, but most are difficult to evaluate. One of the easier combinations requires measurement of the principal angle $\theta_{i'}$, and of the corresponding reflectances of the s and p light. This results in two approximate relationships which can be used to evaluate the optical constants (Jenkins and White, 1976, p. 537).

$$K = \tan\left(2\sqrt{R_{p'}/R_{s'}}\right) \qquad (5.9)$$

$$n\sqrt{1 + K^2} = \sin^2\theta_{i'}/\cos\theta_{i'} \qquad (5.10)$$

The values $R_{p'}$ and $R_{s'}$ are simply the p and s reflectances measured at $\theta_{i'}$.

The relationships presented above, based on the Fresnel equations, are standard fare in many basic texts on optics. Although the dependence of scatter on polarization is somewhat more complicated than the specular relationships discussed above, it should come as no surprise that scatter effects can be explained through the use of the complex dielectric constant (or index of refraction). The next section relates polarization effects in the light scattered from optical surfaces through the use of the polarization factor introduced in Chap. 3.

5.2   The Polarization Factor Q

As indicated in the Chap. 3 discussion of the Rayleigh-Rice perturbation diffraction theory, the polarization factor Q is a real number that relates the effect of surface material properties (as opposed to surface shape) on the BRDF. Its value depends on the sample dielectric constant, as well as the incident angle and the scatter angles. In addition, it depends on the incident polarization and on the polarization(s) allowed to pass to the detector. As one would expect, it accounts for several familiar effects. Evaluated in the specular direction, the expressions for Q reduce to the Fresnel reflectance equations defined in the last section. Brewster's angle, ellipsometry effects, and the presence of plasmons (surface waves) are also accounted for within Q.

Barrick (1970), Maradudin and Mills (1975), and Church (Church and Zavada, 1975; Church et al., 1977, 1979) are responsible for introducing these relationships into the modern radar and optical literature. Q is actually the sum of as many as four different quantities that correspond to the four possible combinations of input and observation polarization. Borrowing Church's notation, the subscripts α and β refer to the incident and observed polarizations, respectively, and Q becomes

$$Q = Q_{\alpha\beta} \tag{5.11}$$

for an α-polarized source and β-sensitive receiver. If the receiver is insensitive to polarization Q becomes:

$$Q = \sum_{\beta} Q_{\alpha\beta} \tag{5.12}$$

for a polarized source

$$Q = \tfrac{1}{2} \sum_{\alpha} \sum_{\beta} Q_{\alpha\beta} \tag{5.13}$$

and for an unpolarized source. The individual expressions for the $Q_{\alpha\beta}$ are

$$Q_{ss} = \left| \frac{(\in - 1) \cos \phi_s}{(\cos \theta_i + \sqrt{\in - \sin^2 \theta_i})(\cos \theta_s + \sqrt{\in - \sin^2 \theta_s})} \right|^2 \tag{5.14}$$

$$Q_{sp} = \left| \frac{(\in - 1) \sqrt{\in - \sin^2 \theta_s} \sin \phi_s}{(\cos \theta_i + \sqrt{\in - \sin^2 \theta_i})(\in \cos \theta_s + \sqrt{\in - \sin^2 \theta_s})} \right|^2 \tag{5.15}$$

$$Q_{ps} = \left| \frac{(\in - 1) \sqrt{\in - \sin^2 \theta_s} \sin \phi_s}{(\in \cos \theta_i + \sqrt{\in - \sin^2 \theta_i})(\cos \theta_s + \sqrt{\in - \sin^2 \theta_s})} \right|^2 \tag{5.16}$$

$$Q_{pp} = \left| \frac{(\in - 1)(\sqrt{\in - \sin^2 \theta_s} \sqrt{\in - \sin^2 \theta_i} \cos \phi_s - \in \sin \theta_i \sin \theta_s)}{(\in \cos \theta_i + \sqrt{\in - \sin^2 \theta_i})(\in \cos \theta_s + \sqrt{\in - \sin^2 \theta_s})} \right|^2 \tag{5.17}$$

Although these equations are fairly intimidating at first glance, and require considerable effort for exact computation (especially when the dielectric constant is complex), examination of several special cases will provide considerable insight into their characteristics. And, in fact, these equations are simply a more general representation of the more familiar relationships presented in Sec. 5.1.

In the plane of incidence ($\phi_s = 0$), the cross-polarization terms (Q_{sp} and Q_{ps}) are zero. In the specular direction, $\theta_s = \theta_i$ and Eqs. (5.14) and (5.17) reduce to the Fresnel reflection coefficients. They are given here for power, instead of field strength, and are expressed slightly differently than Eqs. (5.2) and (5.3).

$$Q_{ss \text{ specular}} = \left| \frac{\cos \theta_i - \sqrt{\in - \sin^2 \theta_i}}{\cos \theta_i + \sqrt{\in - \sin^2 \theta_i}} \right|^2 = R_s(\theta_i) \qquad (5.18)$$

$$Q_{pp \text{ specular}} = \left| \frac{\in \cos \theta_i - \sqrt{\in - \sin^2 \theta_i}}{\in \cos \theta_i + \sqrt{\in - \sin^2 \theta_i}} \right|^2 = R_p(\theta_i) \qquad (5.19)$$

The results of Eq. (5.7) are found if $\theta_i = 0$ is substituted into either Eq. (5.18) or Eq. (5.19). Brewster's law can be derived directly from Eq. (5.19) by setting the numerator equal to zero. Remember that the relative dielectric constant is the square of the refractive index.

As indicated in Chap. 4, the value of Q is necessary to compute the PSD of a reflective surface. The problem is that Eqs. (5.14) to (5.17) are not easy to evaluate without computer help, even if the optical constants are known for the sample in question. Combination of Eqs. (5.14) and (5.18) proves the following identity for the incident plane. This is a very convenient way to compute exact values of Q_{ss} from experimental reflectance data without knowing the sample optical constants (Church, 1989).

$$Q_{ss}(\phi_s = 0) = [R_s(\theta_i) R_s(\theta_s)]^{1/2} \qquad (5.20)$$

Here $R_s(\theta_i)$ and $R_s(\theta_s)$ are the specular reflectances measured at θ_i and θ_s, respectively. For in-plane measurements of good reflectors, this relationship makes exact data analysis much easier. The shape of Q_{ss} for $\phi_s = 0$ is seen from Eq. (5.20) to be much like that of $R_s(\theta_s)$—as shown in Fig. 5.5. Because Q_{ss} is a smooth function, an excellent curve fit may be obtained by measuring sample reflectance at just a few angles of incidence. Further, if $R_s(\theta_i)$ is reasonably large then,

$$Q_{ss}(\phi_s = 0) \simeq R_s(\theta_i) \qquad (5.21)$$

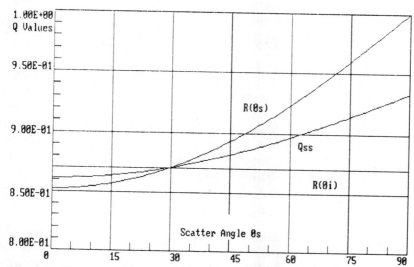

Figure 5.5 Comparison of Q_{ss} with $R(\theta_i)$ and $R(\theta_s)$ for a high-reflectance mirror. The Q_{ss} scale does not start at zero.

which is even easier to use and for most metallic reflectors does not introduce significant error. The point here is that if conversion of the BRDF to the surface PSD is one of the desired results of a scatter measurement, then both source and detector polarization are issues and an s-polarized source with incident plane detection makes the data far more convenient to analyze.

For good reflectors, the absolute value of the dielectric coefficient is much larger than the sin θ's and approximately cancels throughout, simplifying Eqs. (5.14) to (5.17) to the following idealized relationships:

$$Q_{ss} = \cos^2 \phi_s \qquad (5.22)$$

$$Q_{sp} = (\sin \phi_s/\cos \theta_s)^2 \qquad (5.23)$$

$$Q_{ps} = (\sin \phi_s/\cos \theta_i)^2 \qquad (5.24)$$

$$Q_{pp} = [(\cos \phi_s - \sin \theta_i \sin \theta_s)/(\cos \theta_i \cos \theta_s)]^2 \qquad (5.25)$$

The incident plane value of Q_{ss} (see Fig. 5.2 for geometry) is now unity. Out of the plane Q_{ss} falls off to zero as ϕ_s is increased to 90 degrees. This is true even if θ_s is very small. This is because light will not propagate in the direction of the s-polarized electric field. Q_{sp} is identically zero on the incident plane for both the perfect reflector and the exact relationships. Except for unbounded values at $\theta_s = 90°$, it is a nonzero, finite number out of the incident plane. Q_{ps} is very similar except it reaches unbounded values only for $\theta_i = 90°$. The divergence of Eq. (5.25), for Q_{pp} at $\theta_s = 90°$ is due to surface wave effects in Eq. (5.17). For a finite dielectric constant [using Eq. (5.17)] the peak comes at a little less than 90 degrees.

The complete polarization state (E_x, E_y, and δ) is not used to obtain surface statistics. To do so would require that the Q *information* be applied to the fields prior to taking absolute squares—which could be a useful exercise. The next section outlines the techniques reported to describe the complete polarization state of scattered waves.

5.3 Scattering Vectors and Matrices

The Q expressions have been used for the case of smooth, clean, front-surface reflectors. Unfortunately there is no well-established field theory analysis that allows the light scattered from samples not meeting these requirements to be related directly to sample properties. The problem of rough-surface scatter analysis ($\sigma \simeq \lambda$) has defied an exact solution, and the Mie theory, explaining particulate scatter, is restricted to scatter from uniform index spheres. There are, however, ways to characterize, or document, the effect of rough and contaminated samples on the scatter polarization. The techniques are very useful for comparing similar samples and as process control monitors. In other words, it may not be easy to define the relationship between scatter and large defect geometry, but under the right polarization conditions those defects are readily detected. The methods involve first defining vectors that describe the polarization state of an EM wave. The incident and scattered waves are written in terms of the vector definition, and then the vectors are linearly related to each other by a sample-dependent matrix.

The simplest of these methods, known as the Jones calculus or the scattering amplitude matrix, is used to relate the complex s- and p-polarized field vectors of the incident and output waves (Shurcliff, 1962, p. 118; Bohren and Huffman, 1983, p. 61; Azzam and Bashara, 1977, p. 67). It is common to omit the time-dependent terms and leave only the relative phase components. Vector component amplitudes are often normalized by the electric field amplitude. The two components of a Jones vector contain a real and an imaginary part. Thus each vector is described by four variables. These amount to E_x, E_y, δ, and the absolute phase of either component at $t = 0$. Although the Jones calculus is very useful for analysis of polarization of specular elements where the phase relationships are well-preserved (or understood), it is impractical to implement in most optical scatter problems because of the difficulty in measuring the relative phase between the incident and scattered amplitude components. Its strength lies in analysis of specular beams through well-characterized optical components, such as retardation plates. Electrical engineers will be familiar with variations on the Jones calculus, which are used to analyze a variety of waveguide and transmission-line problems.

The Stokes vectors, which are another common way of characterizing the polarization state of an EM wave, are defined in terms of the three critical polarization parameters, E_x, E_y, and δ, as shown in the first half of the following equations (Shurcliff, 1962, p. 18; Bohren and Huffman, 1983, p. 46; Azzam and Bashara, 1977, p. 59). They do not use an absolute phase variable.

$$I = E_x^2 + E_y^2 = \frac{2\eta_0}{A} (P_x + P_y) \tag{5.26}$$

$$M = E_x^2 - E_y^2 = \frac{2\eta_0}{A} (P_x - P_y) \tag{5.27}$$

$$C = 2E_xE_y \cos \delta = \frac{2\eta_0}{A} (P_+ - P_-) \tag{5.28}$$

$$S = 2E_xE_y \sin \delta = \frac{2\eta_0}{A} (P_R - P_L) \tag{5.29}$$

The second half of each of these equations indicates how these vectors may be measured for an electromagnetic wave of time average power $P = P_x + P_y$ over an aperture of area A. The relationship between average wave power and field strength is given in Eq. (A.14) of App. A. M is found from the difference of powers associated with the x and y components. This measurement can be accomplished by using a polarizer in front of the radiometer. C is proportional to the difference in powers $(P_+ - P_-)$ measured by orienting the polarizer at $+45$ degrees and -45 degrees, respectively, from the E_x direction. The last Stokes vector is proportional to the difference in powers $(P_R - P_L)$ found by measuring the right- and left-hand circular-polarized components. This measurement requires a quarter wave plate and a polarizer in front of the radiometer. These four quantities, which overdefine the three-parameter description, are related to each other as

$$(M^2 + C^2 + S^2)/I^2 = 1 \tag{5.30}$$

if the wave is monochromatic and well-polarized. If there is no well-defined phase relationship δ between E_x and E_y (the light is unpolarized), the ratio is zero. Partially polarized light gives a ratio between zero and one. This is the reason for *overdefining* the polarization with four Stokes parameters instead of three. If the light is quasimonochromatic, the four parameters are defined in terms of their time averages and the ratio of Eq. (5.30) will be less than one. Although there are no truly monochromatic light sources, lasers provide a close enough approximation to the situation of Fig. 5.1.

Thus for polarized light, the Stokes vector gives the entire intensity and polarization description of the EM wave. It contains the information necessary to determine δ, the phase between the s and p components, but does not give any information about the absolute phase of the composite wave amplitude. The Stokes vectors for vertically polarized and right-hand circular polarized light are shown below. The vectors can be conveniently normalized by the first parameter I so that each parameter varies between zero and one depending on the polarization state.

$$
\begin{bmatrix} I \\ M \\ C \\ S \end{bmatrix}_V = E_y^2 \begin{bmatrix} 1 \\ -1 \\ 0 \\ 0 \end{bmatrix} \qquad \begin{bmatrix} I \\ M \\ C \\ S \end{bmatrix}_{RC} = (E_x^2 + E_y^2) \begin{bmatrix} 1 \\ 0 \\ 0 \\ 1 \end{bmatrix} \qquad (5.31)
$$

The four-by-four Mueller matrix $[m_{ij}]$ is used with the Stokes vectors to represent the effect of a sample on the intensity and polarization properties of an incident EM wave. The 16 matrix elements are the values necessary to convert an input Stokes vector to an output Stokes vector. The matrix elements change with wave direction, as well as sample properties, and can be used to describe effects induced on transmitted, reflected, and scattered light. As an example, consider the conversion of linearly polarized light from vertical to horizontal by a half-wave plate oriented with its fast axis at 45 degrees from horizontal. By inspection the following matrix correctly performs the conversion. You can quickly confirm that, as expected, it will also reverse the process. Normalized vectors are used.

$$
\begin{bmatrix} 1 \\ 1 \\ 0 \\ 0 \end{bmatrix} = \begin{bmatrix} 1 & 1 & 0 & 0 \\ 0 & -1 & 0 & 0 \\ 0 & 0 & 1 & 0 \\ 0 & 0 & 0 & 1 \end{bmatrix} \begin{bmatrix} 1 \\ -1 \\ 0 \\ 0 \end{bmatrix} \qquad (5.32)
$$

The matrix elements are not derived from field theory, but are empirically found to work, either by inspection or experiment. A Mueller matrix can be found that will convert between any two arbitrary

Stokes vectors; however, not all Mueller matrices are physically realizable. Shurcliff reviews the use of Stokes vectors and Mueller matrices and includes a number of commonly used matrices in an appendix.

The Mueller matrix approach has been applied to scattering problems. Bohren and Huffman (1983) review the methodology and apply the technique to scattering by small particles. Hunt (1973) and Bickel et al. (1976) have reported a scatterometer capable of measuring the various Mueller elements. The technique has been used to measure Mueller elements associated with a variety of particles, fibers, biological samples, and optical reflectors (Bell and Bickel, 1981; Bickel et al., 1986, 1987; Zito and Bickel, 1986; Iafelice and Bickel, 1987).

The power of the matrix approach to characterizing polarization changes is in reducing complex problems to a standard procedure. Set up the situation, turn the crank, and out pops a correct answer. Once a matrix is evaluated it can be used to find the output vector for any input vector of interest. The Stokes/Mueller approach is superior to the Jones calculus for scatter problems because it can handle unpolarized light and avoids the issue of absolute phase. Light diffracted from an isolated spatial frequency can be treated much like a specular reflection because it has a well-defined polarization state. Light scattered from a rough surface composed of many surface frequencies presents a phase front that can vary dramatically with angle, even if a well-polarized source is used. The speckle pattern formed when laser light is reflected from a rough surface is an example. If the measurement aperture accepts many speckles, at best an average relative phase can be defined. Thus the detailed surface information available in the specular reflection (optical constants from ellipsometry, etc.) is not necessarily present. The Stokes-Mueller approach provides a way to characterize this type of sample. There are some serious problems with use of the Stokes-Mueller approach for scatter characterization. For many reflectors only a fraction of the 16 matrix elements will be unique. Measurement of the Mueller elements is not straightforward, and once they are found the elements are not easily related to more conventional sample parameters, such as the optical constants. A new matrix must be evaluated for each desired pair of incident/scatter directions. These difficulties can be reduced by choice of an input vector that forces many of the matrix elements to zero and by the use of automated instrumentation that eliminates many of the calculation problems.

There is no need to be constrained to the historical vector/matrix choices for some problems. For many applications, it is probably enough to know the BRDF associated with the four input/output polarization combinations. An approach that may prove more useful to modern scatter problems is to define wave-vector parameters that result in matrix elements that are more easily used. As an example, the incident vector might be composed of the s and p powers P_{is} and P_{ip}. The scattered vector could be the s and p power/solid angle P_{os}/Ω and P_{op}/Ω. The 2×2 scattering matrix S_{ij}, relating the two vectors, contains four independent elements (S_{11}, S_{12}, S_{21}, S_{22}) that are easily identified as the cosine-corrected BSDF values for the four input/output polarization combinations (ss, ps, pp, sp).

$$\begin{bmatrix} P_{os}/\Omega \\ P_{op}/\Omega \end{bmatrix} = \begin{bmatrix} S_{11} & S_{12} \\ S_{21} & S_{22} \end{bmatrix} \begin{bmatrix} P_{is} \\ P_{ip} \end{bmatrix} \tag{5.33}$$

$$P_{os}/\Omega = S_{11}P_{is} + S_{12}P_{ip} \tag{5.34}$$

$$P_{op}/\Omega = S_{21}P_{is} + S_{22}P_{ip} \tag{5.35}$$

Evaluation of the elements can be accomplished by automating the incident beam polarizer and receiver analyzer positions. In principal, a system like this could operate as an ellipsometer in the specular direction and a polarization-sensitive scatterometer elsewhere.

The use of polarization-sensitive measurements should prove to be a very sensitive tool for defect detection and process-control applications. The key to exploiting these techniques is a firm handshake between the choice of scattering vectors and the instrumentation design. If the polarization vectors are chosen in such a way that the matrix elements are easy to evaluate and relate to physical quantities of meaning, the result will be effective, efficient quality control instrumentation capable of significantly reducing scrap and increasing throughput.

5.4 Summary

A complete description of an EM wave includes its polarization state, as well as its direction, wavelength, amplitude, and absolute phase. The polarization state is defined if the electric-field amplitudes of the s- and p-wave components and the phase difference between them are known. Measurement of the input and output polarization states, along with the other specular beam parameters, allows the sample optical constants to be evaluated through a process known as ellipsometry. The polarization factor Q helps describe the dependence of scatter from smooth, clean, reflective surfaces. It can be evaluated exactly in terms of the complex dielectric constant for the four incident/scattered combinations—ss, sp, ps, and pp. Rough, or contaminated, or volume scattering surfaces cannot be evaluated with these relationships. Instead, scatter is characterized in terms of polarization wave vectors that are acted upon by sample-dependent matrices. The matrix elements are generally found empirically and have no well-defined relationship to material constants.

Polarization-sensitive scatter measurements have the potential for providing a new class of instrumentation for inspection and process control applications both in and out of the optics industry. For a given sample, product, or process, an unwanted defect will often have a matrix element in some preferred direction that is orders of magnitude different than the undamaged substrate, or host, material. By measuring just that particular element, fast, sensitive measurement systems can be developed. Examples are given in Chap. 9.

6

Scatter Measurements and Instrumentation

This chapter reviews the processes and equipment for taking scatter measurements. As pointed out in Sec. 1.5, the BSDF is defined in differential form, but is measured with the incremental limitations imposed by real instrumentation. All of the following produce noticeable deviations between the true BSDF and the measured BSDF: the finite detector aperture, scatter created within the instrument, calibration inaccuracies and practical equipment limitations, such as noise, detector nonlinearity, and mechanical errors. In order to generate meaningful scatter specifications and fully utilize the data, it is important to understand these deviations. Chapter 1 provides necessary background information for understanding this chapter.

6.1 Scatterometer Components

The simple plane-of-incidence scatterometer outlined in Fig. 6.1 contains most of the components typically found in more sophisticated systems. These may be grouped into four categories: light source, sample mount, receiver (detector), and computer/electronics package. This section outlines the need for, and general operation of, these modules.

The light source is formed by a laser beam that is chopped, spatially filtered, expanded, and finally brought to a focus on the detector path. The beam is chopped to reduce both optical and electronic noise. This is accomplished through the use of lock-in detection in the electronics package which suppresses all signals except those at the chopping frequency. The reference detector is used to allow the computer to ratio out laser power fluctuations and, in some cases, to provide the necessary timing signal to the lock-in electronics. In Fig. 6.1 the reference signal is obtained by measuring the light scattered off the chopper

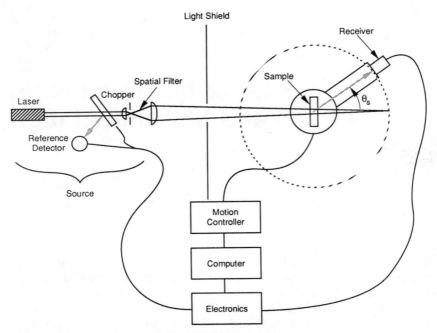

Figure 6.1 Components of a typical BSDF scatterometer.

blade when the beam is blocked. Polarizers, wave plates, and neutral density filters are also commonly found at this location when required in the source optics. The spatial filter removes scatter from the laser beam and presents a near point source which is imaged by the final focusing element to the detector zero position. Although a lens is shown in Fig. 6.1, the use of a mirror, which works over a larger range of wavelengths and generally scatters less light, is very common. The spot size on the sample is obviously determined by elements of the system geometry and can be conveniently adjusted by changing the focal length of the first (cheaper) lens. The source region is completed by a shield that isolates stray laser light from the detector.

The sample mount can be very simple or very complex. In principal, 6 degrees of mechanical freedom are required to fully adjust the sample. Three translational degrees of freedom allow the sample area (or volume) of interest to be positioned at the detector rotation axis and illuminated by the source. Three rotational degrees of freedom allow the sample to be adjusted for angle of incidence, out-of-plane tilt and rotation about sample normal. The order in which these stages are mounted affects the ease of use and cost of the sample holder. In practice, it may prove convenient to eliminate (or occasionally duplicate) some of these degrees of freedom in the sample mount. Exact requirements for these stages differ depending on whether the sample is reflective or transmissive, as well as depending on size and shape. In addition, some of these axes may be motorized to allow the sample area to be raster scanned, to automate sample alignment or to measure reference samples. The order in which these stages are mounted affects the ease of sample alignment.

The receiver-rotation stage is motorized and under computer control so that the input aperture may be placed at any position on the observation circle (indicated by the dotted line in Fig. 6.1). Data scans may be initiated at any location. Systems vary as to whether data points are taken with the receiver stopped or *on the fly*, but unlike the TIS system, the detector is always normal to the incoming scatter signal. Receiver design varies, but changeable apertures, bandpass filters, polarizers, lenses, and field stops are often positioned in front of the detector element. In addition to the indicated axis of rotation, some mechanical freedom is required to assure that the receiver is at the correct height and pointed at the illuminated sample. Sensitivity, low noise, linearity, and dynamic range are the important issues in choosing a detector element and designing the receiver housing.

The mechanical structure allowing the relative positioning of source, sample, and receiver is called a goniometer. Other configurations are possible. For example, the source and sample can be rotated as a unit in front of a fixed receiver (Orazio et al., 1982), or for reflective samples, the scatter pattern can be moved past a fixed receiver by rotating the sample with the source fixed (Church et al., 1977). There is a practical difficulty with configurations that move the sample and/or the source during the measurement. If the sample is optically smooth then the reflected beam sweeps about the lab during the measurement. Unless a moving beam dump is designed to track the beam, the result is a potential safety problem and a lot of unwanted stray light that can become confused with sample scatter (see the next section on instrument signature).

Although less visible, the electronics/computer package represents over half of the effort that goes into a well-designed system. Instrument versatility, throughput, and ease of use are all determined by the decisions made during the design of these elements. Detected signals are likely to vary over a range of 9-to-12 orders of magnitude so some form of data compression or signal processing must be employed to get the signal through an 8-to-16 bit port and into the computer. This usually involves the use of automated electronic gain changes, filter changes, aperture changes, a log conversion device, or all four. In more sophisticated systems the computer controls the data-taking process under the direction of previously entered operator instructions. For example, the operator might ask for data points to be taken every 0.5 degrees over the range 2 to 85 degrees from specular. The aperture size, distance from sample to detector, and total source power (P_i), would also be entered so that the BSDF can be calculated. The specular zero location (position of the focused spot) has to be determined prior to taking data. It is helpful if the computer displays the BSDF during the measurement process.

The ability to store, analyze, and display data in convenient graphical format is a key feature of the instrument. An example of BRDF data from a front-surface mirror is shown in Fig. 6.2. The horizontal axis is a log scale of degrees from specular ($\theta_s - \theta_i$) and the vertical axis is a log scale of the BRDF in sr^{-1}. Notice that the BRDF drops by over 11 orders of magnitude. Measurements are typically in the range 10^7 to 10^{-7} sr^{-1}. The second plot in Fig. 6.2, labeled *instrument signature*, is a measure of light scattered by the instrument and is clearly a concern for BSDF interpretation in the near-specular region. The next section discusses the causes, measurement, and interpretation of instrument signature.

6.2 Instrument Signature

The instrument signature is a combination of several factors. The signature data of Fig. 6.2 was measured by removing the sample and

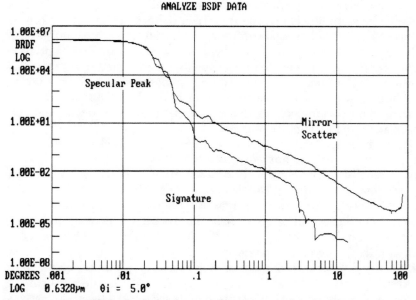

Figure 6.2 The BRDF of a molybdenum mirror compared to instrument signature.

then scanning the incident beam after it passed through the empty sample holder. It is plotted on a log-log scale to emphasize the near-specular region where signature light is usually strongest. The signature profile differs from that of an ideal diffraction-limited spot for three reasons: Light scattered by the instrument is included in the measured profile; the spot contains aberrations caused by the focusing element and is not truly diffraction limited; and as discussed in Chap. 1, the measured profile is broadened because it is actually the result of a convolution of the finite detector aperture with the focused spot. The first two effects are discussed here and convolution broadening, which affects sample scatter as well as signature, is discussed in Sec. 6.3.

Although both stray laser light and room light contribute to signature, the laser is a more severe problem because it is chopped and will easily pass through the detector filter and electronics. Figure 6.3 shows where stray laser light is generated and how it mixes with the measured scatter signal. The dotted line represents scatter signal from the sample that at some value of θ_s will pass through the receiver aperture to become signal. The wavy lines represent stray laser light within the system that can potentially contribute to instrument signature. The focusing optic is a source of scatter and its own BSDF will be added to the sample scatter. However, this signature contribution is reduced, relative to sample scatter, because of the longer distance to the receiver. Once the aperture is moved just off the focused spot the detector face becomes a scatter source illuminating the hemisphere in front of the detector. Every object within the detector field of view (FOV) is illuminated and becomes a source of instrument signature. Of particular importance are the sample holder and the system output-spatial filter. The latter will have the focused spot imaged onto it by the focusing optic. Methods of reducing signature include beam dumps, limiting the detector FOV, and using black surfaces at every possible location (including the output-spatial filter). These issues will

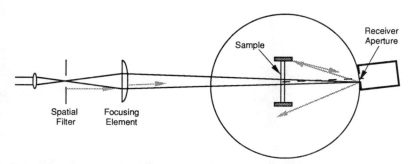

Figure 6.3   Stray scattered light.

be discussed further in the section on near-specular measurements (Sec. 6.4).

The second source of signature is broadening due to aberrations induced by the focusing element. Ideally, this effect can be eliminated by using diffraction-limited optics. However, because spherical mirrors and lenses are usually lower scatter than aspherical reflectors or multielement lenses, the cheaper elements often produce better signatures. This means minimizing aberrations through system geometry is often a design issue. Although aberrations can be predicted from geometrical considerations, and as such are independent of wavelength, the width of a diffraction-limited spot increases with wavelength so the effect of aberrations is lower in the infrared than in the visible. Figure 6.4a shows a signature scan taken at 0.6328 μm with a

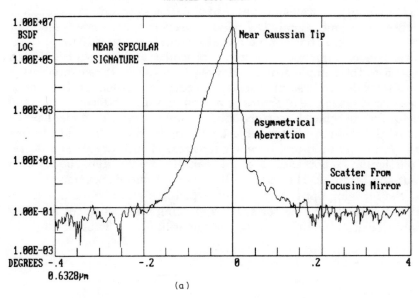

(b)

Figure 6.4 Instrument signature and aberration: (a) Near-specular signature contributions from aberrations; (b) Comatic and spherical aberration from a mirror.

50-cm radius spherical mirror used to focus the spot. The spot, shown in Figure 6.4b, is asymmetrical because of comatic and spherical aberrations introduced by the mirror. This effect, which can also be seen in the near-specular data of Figure 6.4a, starts to become apparent less than one order of magnitude below the peak. If very-near-specular measurements are required, they should be taken on the low-aberration side of specular. At about 0.1 degrees from center, the signature is dominated by scatter from the focusing mirror and this continues until the mirror is out of the detector FOV at about 2 degrees from specular (see Fig. 6.2). This location, which has been dubbed θ_N in the literature (Stover et al., 1987a), will be calculated from system geometry in Sec. 6.4. Beyond θ_N, the signature is dominated by the electronic noise floor. On BSDF plots the noise floor will slowly rise at higher angles as $\cos \theta_s$ decreases. The electronic noise-equivalent BSDF (sometimes NEBSDF) is discussed in Sec. 6.5.

6.3 Aperture Effects on the Measured BSDF

The true BSDF is a complex three-dimensional intensity pattern. Not only does its intensity generally vary by several orders of magnitude, but it often contains a great deal of structure. As we have seen, this structure is directly related to the surface or bulk defects under investigation, so it is important to understand any deviations between the actual BSDF and the measured results. As indicated in Sec. 1.5 and implied by the approximation sign in Eq. (1.8), there is an inherent error in the measurement process. This error, known as aperture convolution, is due to the (necessary) finite size of the receiver aperture.

The effect is easily demonstrated near specular as shown in Fig. 6.5. Here four measurements were made of the same focused spot using circular receiver apertures of four different diameters. The results are dramatically different. The peak value decreases by two orders of magnitude as the aperture diameter is increased from 100 to 2540 μm. The width at the half-power points does just the opposite. Which one is correct? In a certain sense all of them, and in a different sense, none of them are correct.

The BSDF at each measurement location is calculated according to Eq. (1.9) which is repeated here as Eq. (6.1) for convenience. The value P_s is treated as the average over Ω_s and the calculated BSDF assigned to the position θ_s. If the aperture is larger than the focused spot and centered on it then P_s is nearly P_i and the BSDF approaches the constant $1/\Omega_s$. Essentially the same value will be obtained for any position θ_s that allows a particular aperture to capture most of the focused specular light. It will be shown below that reducing the aperture to

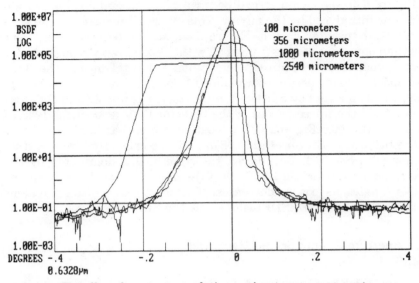

Figure 6.5 The effect of aperture convolution on signature measurements.

zero does not result in an infinite measured BSDF. The distance from aperture to sample was $R = 51$ cm for these measurements which, as graphed, give the value $1/\Omega_s = R^2/\pi r^2 = 12,839$ for the 2540 μm aperture. As the aperture diameter is decreased, the flat section gets shorter. At 100 μm the aperture is about the size of the focused spot. The 100-μm aperture comes closest to the actual BSDF, although even it is in error. If one treats the plots as the convolutions of the apertures with the BSDF, then all four plots are correct. A good approximation is that the measured specular beam width, at the half-power points, is equal to the true half-power spot diameter plus the aperture diameter.

$$\text{BSDF} = \frac{P_s/\Omega_s}{P_i \cos \theta_s} \tag{6.1}$$

Beyond a few tenths of a degree the curves converge, although the 100-μm aperture data have a lot more structure associated with it. The structure is actually there. It is just averaged out in the larger aperture measurements. In order to see it, a small aperture must be used. The data were taken by moving the receiver in one-third aperture steps. The price paid for resolving the structure is additional time and data. Notice that all of the plots in Fig. 6.5 exhibit the aberration-induced asymmetry of Fig. 6.4. The slopes of the four curves, as the spot leaves the aperture, are almost identical. And, if the four curves were integrated, approximately the same result would be achieved for each.

The advantage of a large aperture in low-light level conditions is that the signal to noise is increased. The practical upper limit is usually imposed by the clear aperture of the optics behind the receiver aperture. Typical maximum aperture sizes are 1-to-5 degrees in diameter (as measured from the sample). If the aperture is increased to cover the entire hemisphere in front of the sample, the measured BSDF is just the total reflectance (specular and diffuse) divided by 2π.

Occasionally, it makes sense to reduce the aperture to very small sizes. This is true for some of the near-specular measurements described in Sec. 6.4. However, normally the smallest aperture is kept larger than the focused specular spot. This is because of some unwanted structure that is often imposed on the scatter pattern. If the scatterometer light source is a laser, there will be a speckle pattern modulating the BSDF that depends on source characteristics, as well as sample characteristics. Speckle diameter will be about one-half of the focused spot diameter. If the aperture is kept large enough to accept several speckles, this effect is averaged out. Minimum apertures are typically 0.01 to 0.1 degrees in diameter.

The maximum measurable BSDF has been found (Schiff, 1988) by analyzing the characteristics of the focused laser spot at the receiver aperture. For a Gaussian beam the magnitude of the electric field in the focused spot is given as

$$|E| = E_0 e^{-r^2/\omega_0^2} \qquad (6.2)$$

where E_0 is the electric field value at beam center, r is the radial distance from beam center, and ω_0 is the beam radius where the field has dropped by e^{-1} from center. The value of ω_0 is given as

$$\omega_0 = \frac{2\lambda R}{\pi D} \qquad (6.3)$$

where R is the detector sweep radius and D is the illuminated sample spot diameter. The measured value of P_s is found by integrating the square of the electric field (divided by twice the impedance of free space) over Ω_s. Applying this process to Eq. (6.1) gives an expression for the maximum measurable value of the BSDF.

$$\text{BSDF}_{\text{max}} = \frac{P_s}{P_i \Omega_s} = \frac{(1 - e^{-2r^2/\omega_0^2})R^2}{\pi r^2} \qquad (6.4)$$

If the detector aperture radius is larger than the beam waist radius ($r > \omega_0$), then the maximum measurable BSDF is approximately the inverse of the solid angle $(1/\Omega_s)$ as previously noted. The maximum measurable value goes up as the detector aperture goes down. A limit is approached for the maximum possible measurable BSDF as the detector aperture radius is reduced to zero.

$$\text{BSDF}_{max} (r = 0) = \frac{2R^2}{\pi\omega_0{}^2} = \frac{\pi D^2}{2\lambda^2} \tag{6.5}$$

This is an interesting result because it gives the maximum measurable BSDF independent of instrument geometry. For a 5-mm spot on the sample at 0.633 μm this is $9.8 \times 10^6\ sr^{-1}$. The values predicted by Eq. (6.5) are difficult to reach experimentally because the focused spot is usually aberration broadened; however, values exceeding 50 percent of this theoretical maximum are not uncommon.

The effects of using small, but finite, apertures to measure the differential BSDF have been shown to vary from rather inconsequential to very significant. These difficulties are especially severe in regions where there are large intensity variations over the receiver aperture. Near-specular measurements, described in the next section, are a good example of BSDF data that should be viewed with these limitations firmly in mind.

6.4   Signature Reduction and Near-Specular Measurements

Reduction of near-specular scatter from optics has become an increasingly important issue. Near-specular scatter is particularly troublesome in systems where high-resolution imaging is required. Modern space and aircraft optics often have to meet these types of requirements in the visible and the infrared. Until the mid 1980s, scatter measurements closer than 1 or 2 degrees from specular were difficult or impossible to obtain because of the uncertainties caused by instrument signature. However, as equipment was developed to the point where sample scatter could be separated from instrument signature, these measurements became possible (Stover et al., 1985; Cady et al., 1988; Klicker, 1988). The techniques involve reduction of signature light through proper system design and reduction of measured specular beamwidth reducing the diameter of both the aperture and the focused specular beam.

This section addresses the issues associated with the causes of instrument signature and suggests several methods to reduce it. For the purposes of this discussion, it is assumed that the input specular beam is focused in the detector aperture plane so that the aperture is in the diffraction far field. Most scatterometers achieve the focused spot by imaging the light from a spatial filter with an objective mirror (or lens) onto the detector aperture plane. These arrangements are shown in Fig. 6.6. There are four possible sources of signature light in such a system. These are (1) light leaving the focused specular beam due to reflections in the lens, (2) stray light scattered from the specular beam at the detector, (3) scatter associated with the focusing optic, and (4) aberrations and diffraction associated with the focusing optic. Without proper treatment, any one of these sources can significantly affect the · ability to measure sample scatter near the specular beam.

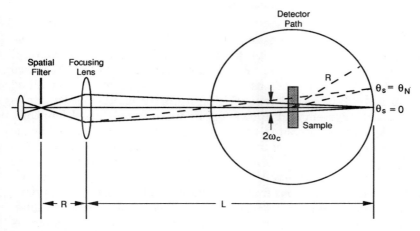

(a) The lens configuration.

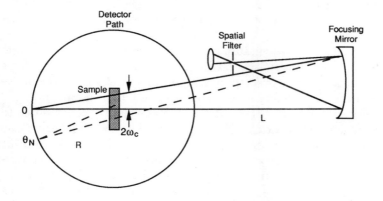

(b) The mirror configuration.

Figure 6.6   Geometical configurations.

Section 6.4.1 examines the choice between using a lens or a mirror to focus the beam, and concludes that a mirror is the better choice. Section 6.4.2 discusses the use of apertures to limit the extent of near-angle scatter. The most effective way to accomplish this is to limit the detector field of view. An expression is found for the maximum angle from specular θ_N over which near-specular signature light is expected to play a significant role. Reduction of signature at observation angles inside θ_N is the topic of Sec. 6.4.3. Light scattered from the focusing element presents the largest problem.

6.4.1 Reflective versus refractive focusing optics

There are two configurations of focusing optics shown in Fig. 6.6. One illustrates the use of a lens, while the other uses a mirror as the final focusing element prior to the sample. Comparisons between these two configurations will be made assuming that the system is constrained to fit on a 4 × 8-ft optical table. Using a mirror for the focusing element folds the system and allows for the use of longer focal lengths.

Of the four contributors to instrument signature, light reflected off the lens surfaces is present only in the refractive system. Reflections off the lens surfaces result in three cones of light superimposed on each other. The first cone is due to a double reflection that proceeds forward to the detector. The other two are reflections from each individual lens surface that propagate in the reverse direction and illuminate the spatial filter. The illuminated region around the spatial filter is imaged in the detector plane creating signature light. These reflective sources of light have been examined in detail through the use of a ray tracing program (Stover et al., 1987a), and have been found to be quite severe for lenses of all shapes. There is no choice but to eliminate it through the use of antireflection coatings on the lens. Unfortunately, these coatings tend to be higher scatter than the surfaces they cover.

Another contribution to signature is stray light scattered into the system when the specular beam reflects off the detector aperture housing. This light illuminates virtually everything within the field of view of the detector. There will be little difference between using a lens or a mirror.

Scatter associated with the focusing element clearly contributes to the instrument signature light. The mirror will scatter light from one surface. The lens will scatter from two surfaces and its bulk. Similar single-surface scatter characteristics are expected for the best spherical metal and dielectric surfaces, so the best lens probably has more scatter than the best mirror. In addition, as mentioned above, the lens must be antireflection coated, which usually increases the scatter. The angular distribution of the scattered light and distance to the detector also play a role, as will be described later.

The fourth source of signature is due to focused spot aberrations and diffraction. Because spherical surfaces can be made lower scatter than aspheres, aberrations are almost a certainty. The specular point source (spatial filter) can be located on the principal axis of the lens so that off-axis aberrations, such as coma and astigmatism, are not present in the refractive system. This is not true for the mirror which must be tilted to separate the incident and reflected beams.

The aberrations associated with the two systems can be compared with the use of spot diagrams produced by a ray-tracing program. The comparison is made with the two configurations constrained to fit on a 4 × 8-ft table. The angle of convergence of the focused beam at the detector is kept the same for each case. This allows the relationship between sample spot size ($2\omega_s$) and sample-to-detector distance (R) to be identical for the two cases. (This is an important consideration, as will be shown in the next section.) The distance from the focusing element to the detector (L) was 150 cm for the lens and 250 cm for the mirror. The two cases have been examined through the use of a ray-tracing program. Spot diagrams, taken in and near the focal plane, are shown for each configuration in Fig. 6.7. Notice that the scales are different for the lens spot diagram. The lens was optimized for minimum spherical aberration (nearly planoconvex); however, even for this situation it produces a considerably larger aberrated spot than the mirror. This is true even though the mirror produces both astigmatism and coma, in addition to spherical aberration. Mirror tilt was minimized to reduce aberrations. The size of the diffraction-limited spot for this geometry is also shown. For a fixed R as $2\omega_s$ decreases, the aberrations will go down and the diffraction limit will increase. This effect is considered further in the next section.

Table 6.1 summarizes the signature comparison of the reflective and refractive systems. The mirror configuration is clearly superior.

6.4.2 Minimizing the near-angle/far-angle boundary, θ_N

The measured contributions to instrument signature from spot aberrations and diffraction can normally be confined to within a few tenths of a degree from specular by choosing a suitably small receiver aperture. The scatter contribution from the focusing optic extends much farther out and determines the boundary angle θ_N beyond which the near-specular contributions to signature are greatly reduced. θ_N is defined as the angle from specular at which the illuminated spot on the focusing element has left the receiver field of view. Examination of Fig. 6.6 makes it clear that when the receiver is close to the specular beam, it can look through (or off) the sample and see scattered light coming from the focusing optic. As the receiver moves

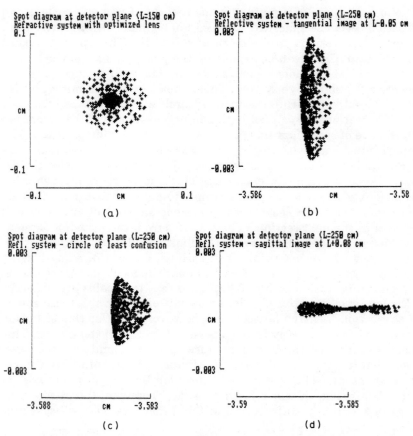

Figure 6.7 The circular focused spot from the lens in (a) is enlarged to about 1 mm diameter by spherical aberration. Off-axis aberrations associated with a focusing mirror cause the spot to change shape along the focus path (b), (c), (d), but the largest dimension is about 0.06 mm, which is considerably smaller than the spot formed by the lens.

TABLE 6.1   Lens/Mirror Comparison

Source	Lens	Mirror
1. Lens reflections	AR elimination	Nonexistent
2. Stray specular light	Equal	Equal
3. Scattered light	Slightly larger	Slightly smaller
4. Aberrations	Large	Small

away from specular, the sample holder (or any other beam aperture) will block this light from reaching the detector and reduce signature. The maximum receiver angle θ_N at which light scattered from the focusing element contributes to the instrument signature, is a function of instrument geometry and may be found from Fig. 6.8, which shows the detector located at θ_N. Scattered light works its way from the top of the focusing element (radius r_m) through the bottom of the sample (radius r_c) and just into the edge of the receiver aperture (radius r_s) whose field of view is limited by the sample holder. The small-angle approximation, $\sin \theta = \theta$, has been made throughout the derivation.

$$\theta_N = (r_m + r_c)/L + (r_c + r_s)/R - r_c/F \qquad (6.6)$$

The last term covers the special case of a sample with focal length F ($F > 0$ for a converging sample) and is derived under the assumption that the source optics are adjusted to bring the specular beam back to focus after the sample is inserted (Klicker et al., 1988). Substituting typical component values of $R = 50$ cm, $L = 100$ cm, $r_m = 5$ cm, $r_c = 2$ cm, $r_s = 0.2$ cm, and $F = \infty$ gives a value of 6.5 degrees for θ_N which is fairly large. This can be reduced by limiting the receiver field of view to an area on the sample a little larger than the illuminated spot and by considering the effective mirror size to be limited to its illuminated portion. Assuming a Gaussian beam of radius ω_c (e^{-2} power point) at the sample and limiting the field of view to a value of $r_c = 2\omega_c$. In similar fashion, the effective mirror radius is chosen as twice the beam radius at the mirror ($r_m = \omega_m$). Because the beam is focused at the receiver the two-beam radii are related as $\Omega_m = \Omega_c(R + L)/R$. Substituting into Eq. (6.6) gives

$$\theta_N = 4\omega_c(1/L + 1/R) + r_s/R \qquad (6.7)$$

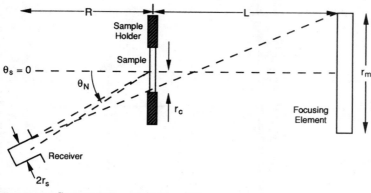

Figure 6.8 Geometry for calculation of θ_N.

which has been presented by other authors after slightly different assumptions (Lee et al., 1986). Using $\omega_c = 0.2$ cm gives $\theta_N = 1.6$ degrees which is much better. After dropping the first and last terms, which are smaller, the result is

$$\theta_N > 4\omega_c/R = \text{the observed source convergence angle} \qquad (6.8)$$

which is a convenient rule of thumb for designing near-specular scatterometers.

Figure 6.9 illustrates a simple method of restricting the field of view to the illuminated spot. A lens behind the receiver aperture images the illuminated sample onto a field stop. The field-stop aperture is then the receiver field-of-view limit and acts to limit θ_N. Restricting the detector field of view in this manner also eliminates the stray light that is reflected off the sample mount. When this receiver system is used, the receiver solid angle is defined by the detector aperture and the field of view does not change as the receiver aperture is increased.

6.4.3   Scatter measurement inside θ_N

Signature contributions inside θ_N can be reduced by obtaining a focusing mirror that is low scatter near specular. This *catch 22* situation (it takes one to measure one) was confronted in the mid-1980s as the first near-specular scatterometers were built (Stover et al., 1985; Cheever et al., 1987). As mirrors get better, so will the scatterometers that measure them. Using low-reflectance materials for the receiver aperture and the source output-spatial filter, will also reduce near-specular signature.

In order to measure near the focused specular beam a small receiver aperture must be used. Very-near-specular measurements may make use of apertures that are about the same diameter as the focused spot. Apertures less than 0.1-degree wide are not uncommon and apertures as small as 0.003 degree have been used. Away from specular, the ap-

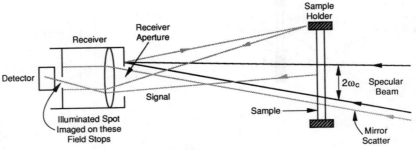

Figure 6.9 Limiting the detector field of view.

ertures must be larger so that the individual speckles, inherent in scatter patterns from laser sources, can be averaged out, and as discussed in Sec. 6.5, so the noise equivalent BSDF can be reduced. Because the measurements can change rapidly near specular, the data points are spaced less than an aperture width apart. To avoid convolution discontinuities, the aperture changes should be made on relatively flat sections of the BSDF. And for the comparison between signature and sample data to be meaningful, the aperture changes should come at the same locations for each scan.

Figure 6.10 shows near-specular scatter data taken on a TeO_2 Bragg cell (Cady et al., 1988) at a wavelength of 0.86 μm. Separation from the instrument signature was achieved at about 0.009 degree from specular. Three scatter regions have been identified on the plot corresponding to signature scatter, bulk scatter, and surface scatter.

Near-specular data from reflectors can be converted to surface statistics using the reasoning of Chaps. 3 and 4. The grating equation, Eq. (1.6), is used to find the associated spatial wavelengths. For near-normal incidence and near-specular scatter the relationship is

$$1/f = \lambda/(\theta_s - \theta_i) \qquad (6.9)$$

which gives a surface wavelength of 4.9 mm at 0.86 μm and 0.01 degree. It would appear from Eq. (6.9) that the maximum observable

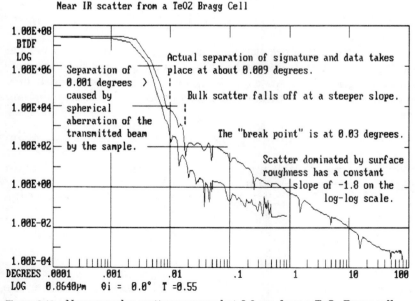

Near IR scatter from a TeO2 Bragg Cell

Figure 6.10 Near-specular scatter measured at 8.6 μm from a TeO_2 Bragg cell.

spatial wavelengths could be increased even further by increasing the source wavelength. However, an increase in source wavelength results in a larger diffraction-limited spot with a resulting increase in the separation angle.

6.5   The Noise-Equivalent BSDF

The minimum measurable BSDF, limited by electronic noise, has also been investigated (Schiff, 1988) and is outlined here. By careful design of the system electronics, it is often possible to reduce the noise to the point where it is dominated by detector noise. For this case, the noise can then be expressed as a function of the detector alone in terms of its noise-equivalent power (NEP). The NEP, or apparent light-noise signal, of the detector can be found from either of the following relations:

$$\text{NEP} = (A_d)^{1/2}/D^* \tag{6.10}$$

$$\text{NEP} = I_n/R_k \tag{6.11}$$

where D^*, A_d, I_n, and R_k are the detector detectivity, active area, noise current, and responsivity, respectively. NEP has units of watts per square root hertz. The equivalent total noise power (P_N) at the detector is found from the NEP by multiplying by the square root of the system noise bandwidth (BW_N).

$$P_N = \text{NEP} \sqrt{BW_N} \tag{6.12}$$

The noise bandwidth of the system is found by approximate numerical analysis techniques (Krauss et al., 1980) and is greater than the normal signal bandwidth. An expression for the noise bandwidth of a lock-in amplifier/digital numerical integration combination (common to these systems) is

$$BW_N = 1.57/(T_I + 2\pi T_C) \tag{6.13}$$

where T_I is the digital integration time and T_C is the lock-in amplifier-output time constant.

The noise-equivalent BSDF can then be expressed in terms of the detector noise P_N, as follows:

$$\text{NEBSDF} = \frac{P_N}{P_i \Omega_s \cos \theta_s} \tag{6.14}$$

The noise-equivalent BSDF can be reduced by increasing either the total incident power (P_i) or the receiver solid angle (Ω_s). Values of about $10s^{-7} sr^{-1}$ can be achieved with a 5 mW HeNe laser and a silicon detector.

Another source of noise that limits the minimum measurable BSDF is scatter from particulates in the air within the illuminated field-of-view volume. The size of the observable illuminated volume increases considerably when the receiver is near specular. In this region the instrument signature is usually dominated by the effects discussed in the preceding sections. At higher angles, however, the noise floor can be determined particulate scatter. The effect is illustrated in Fig. 6.11, where three signatures are compared. A high-efficiency particulate air (HEPA) filter was suspended over the sample area. The three signatures were obtained with the filter off, with the filter on, and with the receiver aperture blocked (to obtain the electronic noise floor). The filter acts to reduce the high-angle signature by about one order down to the electronic noise floor (which is very near the theoretical minimum for the room-temperature silicon detector used). The value of θ_N was 3 degrees and the wavelength was 0.633 μm. The near-specular asymmetry is due to the filter being off-center above the sample. Filtering over the entire scatterometer would further reduce the signature from 3 degrees out to about 10 degrees. The angle-sensitive effect is clearly revealed if one remembers that 100 minus 10 and 10 minus 1 correspond to equal distances on the log BSDF scale. Scatter from

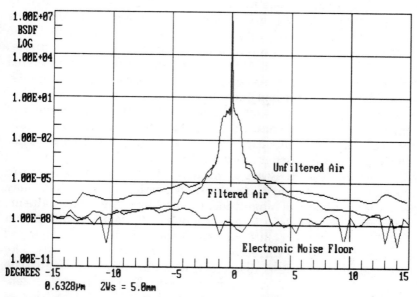

Figure 6.11 Comparison of instrument signature with and without air filtering to the electronic noise floor.

small spherical particulates is proportional to the fourth power of the inverse wavelength and to the sixth power of particulate diameter, so the effect is far more pronounced in the ultraviolet and mildly larger particulates. If a photomultiplier is used in the receiver, theoretical noise floors can drop below $10^{-10} \, sr^{-1}$ and clean air becomes very important.

It is important to realize that the noise signal is statistical in nature. This can be seen by examining the signature of Fig. 6.2 at high angles. The signal beyond a few degrees is due to random electronic noise fluctuations and will not repeat if the signature is remeasured. Therefore, its level should be expressed as an rms value. The rms value of such signals is defined just as it was for surface topographies in Eq. (2.3). The rms value of a random noise signal can be estimated with reasonable accuracy as one-third of the peak signal value.

6.6   Measurement of P_i and Instrument Calibration

Regardless of the type of BSDF measurement, the degree of confidence in the results is determined by instrument calibration, as well as by attention to the measurement limitations previously discussed. Scatter measurements are often received with considerable skepticism. In part, this has been due to misunderstanding of the definition of BSDF and confusion about various measurement subtleties, such as instrument signature or aperture convolution. However, quite often the measurements have been wrong and the skepticism is justified. The results of several round-robin measurement comparison studies (Young, 1975; Leonard, 1988, 1989) attest to the difficulties associated with these measurements. As with many new metrology techniques, these problems are overcome once they are understood.

Instrument calibration is often confused with the measurement of P_i, which is why these topics are covered in the same section. To understand the source of this confusion, it is necessary to first consider the various quantities that need to be measured to calculate the BSDF. From Eq. (6.1), they are P_s, θ_s, Ω_s, and P_i. The first two require measurement over a range of values. In particular, P_s, which may vary over many orders of magnitude, is a problem. And in fact, linearity of the receiver to obtain a correct value of P_s, is a key calibration issue. Notice that an absolute measurement of P_s is not required, as long as the P_s/P_i ratio is correctly evaluated. P_i and Ω_s generally take on only one (or just a few) discrete values during a data scan. The value of Ω_s is determined by system geometry. The value of P_i is generally measured in one of two convenient ways.

The first technique, sometimes referred to as *the absolute method* makes use of the scatter detector (and sometimes a neutral density filter) to directly measure the power incident upon the sample. This method relies on receiver linearity (as does the overall calibration of BSDF) and on filter accuracy when one is used. The second technique, sometimes referred to as *the reference method* makes use of a known BSDF reference sample (usually a diffuse reflector and unfortunately often referred to as the *calibration sample*) to obtain the value of P_i. The reference sample scatter is measured and the result used to infer the value of P_i from Eq. (6.1). This method depends on knowing the absolute BSDF of the reference. The choice of methods is usually determined by whether it is more convenient to measure the BSDF of a reference or the total power P_i. Both are equally valid methods of obtaining P_i. Neither method constitutes a system calibration, because calibration issues such as an error analysis and a linearity check over a wide range of scatter values are not addressed over the full range of BSDF angles and powers when P_i is measured. The use of a reference sample is an excellent system check.

System calibration requires an error analysis on the four quantities in question. In order to accomplish this for P_s, the receiver transfer characteristic, signal out as a function of light in, must be obtained and checked for linearity. This may be obtained through the use of a known set of neutral density filters or through the use of a comparison technique that makes use of two data scans—with and without a single filter (Cady, 1989*b*). Section 6.12 outlines an error analysis for BSDF systems. Full calibration is not required on a daily basis. Sudden changes in instrument signature are an indication of possible calibration problems. Measurement of a reference sample that varies over several orders of magnitude is a good system check. It is prudent to take such a reference scan with data sets in case the validity of the data is questioned at a later date. A diffuse sample, with nearly constant BRDF, is a good reference choice for the measurement of P_i but a poor one for checking system calibration.

Wavelength and angle-of-incidence scaling can be excellent checks of instrument calibration. In order for the check to work, the scaling properties of a sample must be known. Topographic scaling of smooth, clean, reflective samples, as described in Chaps. 4 and 5, is the easiest check to make. Not all front-surface mirrors will scale topographically, because of other sources of scatter (Stover et al., 1989). Many thinly coated optics do not scale and beryllium mirrors do not scale. Solid molybdenum mirrors have been shown to scale from the blue to the near-IR. Aluminum and copper have shown scaling. Gold and silver would be expected to scale but are a poor choice for a reference mirror because they are soft and may damage with cleaning. A silicon carbide mirror that scales would be an excellent choice because its hardness resists damage, but they appear to scatter much like beryllium.

An ASTM standard on BRDF measurements is expected to be completed in 1990 or 1991. In draft form it calls for essentially the same calibration checks as those outlined above. Its publication is a reflection of government and industry policies that will require more formal assurances of accuracy in both instrumentation and measurements.

6.7 Measurement of Curved Optics

Measurement of scatter from curved optics presents a new set of problems. In order to bring the beam back into focus at the receiver path, after inserting the sample, the source optics must be adjusted. This can be accomplished by moving the lens/spatial filter combination shown to the left of the focusing element in Fig. 6.3. This point source of light is conjugate with the focused spot at the receiver. The direction of motion depends on whether the sample is converging or diverging. For samples with short focal lengths, the source may need to be focused very close to the sample which can drastically reduce sample spot size.

The problems are more severe for near-specular measurements, because when the source is adjusted the instrument signature changes. Fortunately, the changes can be accounted for with computer modeling if necessary (Klicker et al., 1988). The diffraction-limited spot decreases in size if the sample is converging and increases if it is diverging. θ_N also decreases for a converging sample, but not by the same amount. The change in θ_N is accounted for in Eq. (6.6).

6.8 Coordinate Systems and Out-of-Plane Measurements

Out-of-plane measurements raise some difficulties that are not always confronted with measurements taken in the incident plane. And there are some sample orientation issues for in-plane measurements that still need to be discussed. In addition to keeping track of the position of source and receiver relative to the illuminated spot (or volume), the angular orientation of the sample and the location of the illuminated spot must be recorded. The geometry used in Fig. 1.6 to define BSDF is redrawn in Fig. 6.12a, showing a reflective sample, with more detail to illustrate these issues. For the moment, the discussion is limited to a flat reflective sample.

Figure 6.12b shows the reflective sample plane with two sets of cartesian coordinates superimposed on it. These are the beam coordinates (xyz), which will be defined by the incident and reflected beams and a set of coordinates (XYZ) fixed to the sample. Thus z and Z are normal to the sample surface and the other four axes are in the sample plane. The $-x$ axis is the projection of the incident beam on the sample plane. This coordinate system moves with the illuminated spot

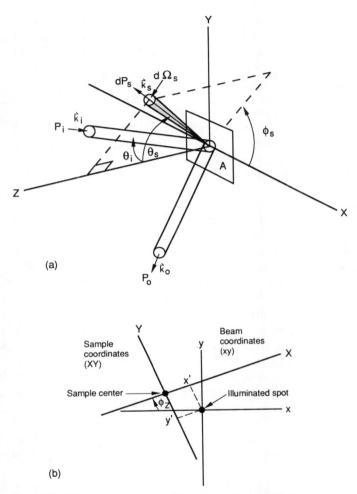

(a)

(b)

Figure 6.12 Scatter angles are defined in terms of the beam coordinates (xy) shown in (a). The (x,y) axes are determined by the incident beam and specular reflection with x in the incident plane. Thus, $\phi_i = \pm 180°$ for all cases. (b) Sample orientation and the illuminated spot location are given in terms of the sample coordinates (X,Y) and a rotation angle ϕ_z. The (X,Y) axes are often indicated by fiducial marks located on the sample edge.

on the sample. The incident and reflected beam directions are given by the propagation vectors k_i and k_0, respectively. These two vectors define the xyz (beam coordinate) system. The XYZ system is used to define *sample center* and sample orientation. The location of the X and Y axes is usually accomplished through the use of fiducial marks on the sample or its holder. These marks, and the XY coordinates of the illuminated spot allow the sample to be measured at known, repeatable locations. For flat samples the Z axis is normal to the sample surface. When the sample is rotated about z or Z, the xy axes are separated in angle from the XY coordinates by the rotation angle ϕ_z, but the Z and z axes remain parallel. The polar and azimuthal angles can now be defined in terms of the axis sets and the propagation vectors.

The polar angles θ_i and θ_s are defined as the angles between z and the propagation vectors k_i and k_s, respectively. The azimuthal angles ϕ_i and ϕ_s are defined as originating from the x axis to the projections of the propagation vectors k_i and k_s on the sample face (XY or xy plane). This definition fixes ϕ_i equal to 180 degrees independent of sample rotation.

Some authors define the azimuthal angles (ϕ_i and ϕ_s) from X instead of x and use this angle as a measure of sample rotation. This is, for example, the choice made by Nicodemus et al. (1977). Unfortunately, with this definition, the value of ϕ_s cannot be substituted into either the grating equation or the expressions for Q without first reducing it by the sample rotation angle. This is because all of those equations assume the special case $\phi_i = 180°$. This can lead to confusion when these equations are used to interpret results. However, there is another problem with this definition that is potentially more serious. If the sample is curved, xy and XY will in general be nonplanar and z and Z will be nonparallel. Now a new set of axes is required or some difficult transformations must be made to correct all the polar and azimuthal angles. All of these complications are avoided if the incident and scattered angles are defined relative to beam coordinates, and the illuminated sample spot location and sample orientation are defined with the sample coordinates. With a beam-coordinate definition the angles are defined in a manner that is independent of sample holder design. Thus definitions (and standards) can be set that are both instrument independent and consistent with equations and notation that currently exist.

When measurements are taken out of the incident plane, the situation can be confusing from another aspect, because the receiver may not move naturally along paths of constant thetas or fees. Figure 6.13 shows the two common ways to cover the full hemisphere in front of the sample. Essentially they may be thought of as viewing lines of longitude and latitude on a globe of the earth from either a position above a pole or in the equatorial plane. The former, shown in Fig. 6.13a, allows the receiver to be moved along paths of constant thetas and fees. This is very convenient because in order to measure just s or p polarization the receiver analyzer can remain at a fixed angle during a scan. However, there are often practical reasons for picking other receiver sweep patterns—such as the one in Fig. 6.13b. These may include sample

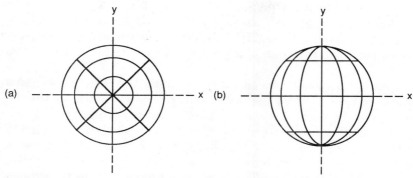

Figure 6.13 Hemispherical receiver paths projected onto the xy plane.

size and/or required field of view (which increase receiver arm length), receiver weight, and multiple source requirements. Instrument configurations that involve tilting the sample about the x axis, so that the beam is reflected out of the initial incident plane and around the lab, are usually avoided because of the resulting increases in instrument signature and safety problems associated with undumped beams.

6.9 Raster Scans

The presence of light scatter from a local area, on an otherwise uniform optic, indicates the presence of a defect or a contamination site. Unfortunately, complete sample coverage by full angle BSDF inspection is often impractical due to time and cost limitations imposed by sample size and/or sample numbers. For these situations a raster scanning technique, which rapidly covers the required sample area, is often the best solution. Raster data provide valuable insights into sample nonuniformity caused by production processes and contamination.

Sample raster scans are accomplished by measuring the BSDF at constant incident and scatter angles from specular over the sample area of interest (Rifkin et al., 1988). This can be done by moving the sample in its own plane with the receiver fixed at one position. The BSDF value, measured at the fixed-receiver angle, is assigned to the corresponding illuminated sample area (or pixel). Raster data are presented either as color maps or isometric *3-D plots* of the scanned sample area. The results reveal the locations of high- and low-scatter areas on the sample. Because scatter patterns are often asymmetrical, the sample must be moved in an x,y grid pattern (as opposed to an r,θ area coverage) so that the pattern is not rotated during the measurement process. Moving the sample is time consuming—approximately 1 pixel/second. At this rate a 10,000 pixel scan takes almost three hours.

Much faster scans can be taken by imaging the fully illuminated sample onto a CCD array. Sample pixels are now defined by the detector array. Because the imaged pixels are not uniformly illuminated, the incident power associated with each pixel must be found prior to calculating the BSDF using Eq. (6.1). Although the values of P_s and P_i change as a function of pixel location, the values of Ω_s and θ_s remain essentially constant. The receiver solid angle Ω_s is determined by the full aperture of the CCD camera. The calculation process is shown in Figs. 6.14 to 6.16 for the case of an incident beam with a near-Gaussian intensity cross section. Color mapping (shown as grey scales here) is used to indicate light intensity. Figure 6.14 is an intensity map of the incident beam obtained by imaging a uniform white diffuse surface. The spot is elliptical because the angle of incidence was not zero. The image of an illuminated front-surface mirror is shown in Fig. 6.15. The data of Figs. 6.14 and 6.15 are used to calculate the sample BRDF, which is mapped in Fig. 6.16. Notice that the prominent scratch across the width of the sample appears relatively uniform in this figure, even though its illumination in Fig. 6.15 is uneven. Resolution on the sample is about 10 μm. The 150,000 pixels

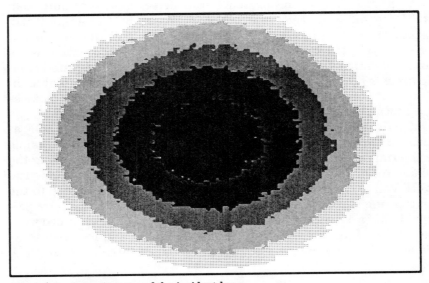

Figure 6.14 Intensity map of the incident beam.

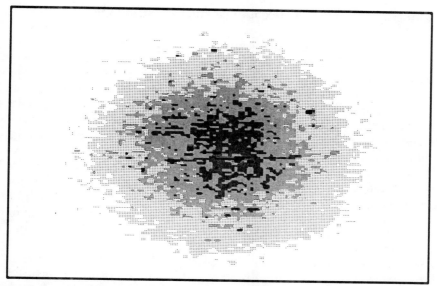

Figure 6.15 Intensity map of the illuminated sample.

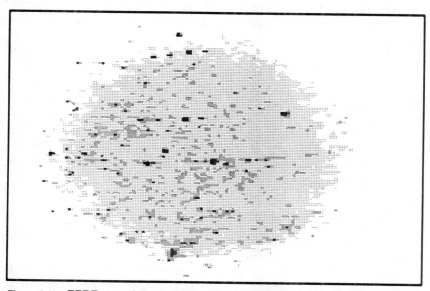

Figure 6.16 BRDF map of the sample resulting from the data of Figs. 6.14 and 6.15.

making up this image were obtained and displayed in just a few minutes.

If an acceptable BSDF level can be established, the inspection time can be reduced further by eliminating the need for a complete color plot of the sample. Instead, the computer can simply check for pixels that exceed the acceptance level and either reject the component or plot the high scatter points, as required. The short measurement time allows easy integration of this inspection technique with other processes, such as various surface finishing methods, dielectric coating, and laser-damage threshold testing.

The advantages of quickly inspecting large numbers of relatively inexpensive mirrors prior to their integration with more expensive system components are obvious. Laser mirrors, requiring a check of only the center spot, can be inspected at a rate of about one per second. Measurements can be made immediately after dielectric coating using the same part holder.

Systems intended to measure large-area optics can also be designed. After determining the source intensity, the sample can be measured one piece at a time by moving the instrument system (source, filters, lenses, and detector) to adjoining fields of view. After all the data are taken, the computer *cuts and pastes* the various pieces together to form a complete picture. For large-area, high-resolution scans, a huge amount of data will have to be stored. However, by retaining only the data associated with defects that exceed operator chosen limits (in size and/or BSDF) this difficulty is eliminated.

This rapid measurement technique is expected to play a significant role in the inspection of both high-volume production optics and large-area optics. Because it provides production-process information in addition to acting as a quality control, it is expected to have a positive impact on several segments of the optical industry. Raster scans are also used in the detection of subsurface defects in both optics and semiconductors as described in Chap. 9.

6.10   Measurement of Retroreflection

Measurement of light scatter back in the incident direction by conventional means is difficult because the receiver shadows the sample from the source. The data shown are for a ring-laser gyro mirror. Minimizing retroreflection, or retroscatter, is important to the correct operation of optical gyros. These devices are ring lasers with cavity modes propagating in the clockwise and counterclockwise directions simultaneously (Fig. 6.17a). If the laser is rotated during operation, the two modes are frequency shifted in opposite directions. One doppler shifts up and the other down. Rotation is detected by combining the two

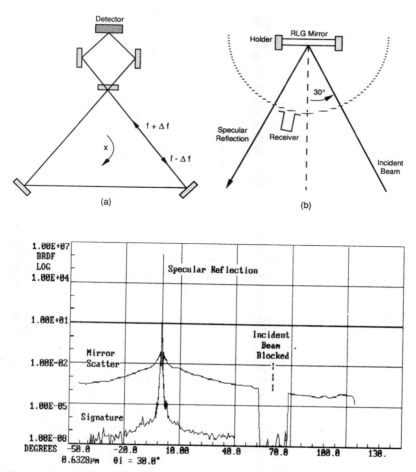

Figure 6.17 (a) A simplified schematic of a ring-laser gyro. (b) Scatter measurement geometry for a ring-laser gyro mirror. (c) Data taken using the arrangement of Fig. 6.17b.

beams outside the laser and watching for motion of the resulting fringe pattern. The resolution with which rotation can be monitored is controlled by the degree to which the frequencies of two counter propagating beams are separated. Retroscatter from the gyro mirrors acts to mix the two beams in the laser cavity and thus limits resolution.

The configuration of Fig. 6.17b was used to obtain the data of Fig. 6.17c. It would appear to be easy to infer the BRDF in the retrodirection from the measured scatter levels on either side. However, some materials have BRDFs that are enhanced in the retrodirection. Referred to as *the opposition effect, enhanced back-*

scatter, or just *retroscatter* the effect takes the form of a narrow peak (approximately one degree wide) that can vary from a few percent to several hundred percent higher than the background scatter levels. The effect is easily observed from airplanes by eye, with sunlight as the source. It is most evident as a halo, appearing around the airplane shadow when flying above clouds. However, it can also be observed as a faint bright spot on the ground (located where the airplane shadow should be) that travels with the plane. The intensity of the spot varies with ground cover. Although it is narrow enough to have little impact on total reflectance, it has direct consequences in the design of laser radar systems and targets, can be of practical importance to air (or space) ground surveillance systems, and as mentioned, is a problem for laser gyros.

Gu et al. (1989*a*, 1989*b*) review the various theoretical explanations that have been advanced to explain the effect. And, in fact, different mechanisms may be responsible for enhanced backscatter from different materials (water droplets, diffuse reflectors, metal mirrors, etc.). Gu et al. (1989*a*, 1989*b*) report a clever doppler shift technique to measure retroscatter directly. The sample is moved, inducing the doppler shift, and its magnitude is determined by monitoring the amplitude of the resulting difference (or beat) frequency after recombining the retrobeam with a reference. Beam splitters can be used to separate retroscatter from the incident beam if the target samples are much higher scatter than the beam splitter and associated beam dumps (see Fig. 6.18). Lenses are used to put the sample in the far field of the source and the detector in the far field of the sample. If a pulsed laser is used as a source and the beam splitter/target distance is suf-

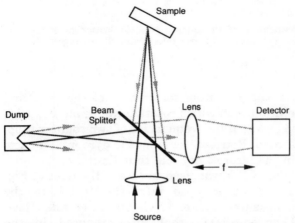

Figure 6.18 A beam splitter used to measure retroscatter from a diffuse sample.

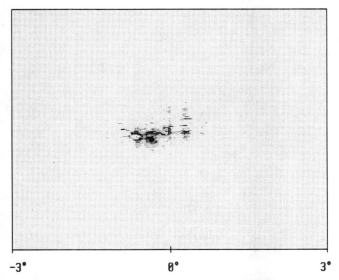

−3°   0°   3°

Figure 6.19 Retroscatter from a white spectralon sample. The spot is about 0.5 degrees wide and about 20 percent brighter than the background.

ficiently large the sample scatter can be separated from that of the beam splitter. If a CCD array is employed as the detector, a map of the scatter through the incident angle can be made as shown in Fig. 6.19, where the sample has been rotated to average speckle effects.

6.11 An Alternate TIS Device

Section 1.6 defines TIS and reviews the Coblentz sphere technique developed for its measurement. Another method is outlined here that may have some advantages over the original technique. The difference is the use of a high-reflectance, diffuse integrating sphere, instead of the specular Coblentz sphere, to gather the scattered light.

The system, shown in Fig. 6.20, consists of a laser source, an integration sphere, three detectors, and associated electronics and optics. The laser beam is divided by a beam splitter, allowing the transmitted power to enter the first detector. The reflected segment becomes the incident power on the sample. The specular reflection from the sample returns to the beam splitter and part of it is transmitted to a second detector. A third detector views the interior of the integration sphere. It is shielded by a baffle from directly viewing the sample. Light scattered out of the specular beam by the sample is trapped in the sphere. The inside of the sphere and the baffle are coated with a diffuse, high-

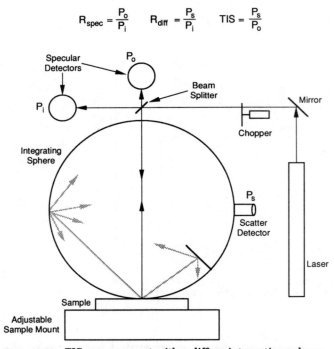

$$R_{spec} = \frac{P_0}{P_i} \qquad R_{diff} = \frac{P_s}{P_i} \qquad TIS = \frac{P_s}{P_0}$$

Figure 6.20   TIS measurement with a diffuse integrating sphere.

reflectance coating that, after multiple reflections, causes the interior of the sphere to be uniformly illuminated. The relative sensitivity of the three detectors and the transmission/reflection ratio of the beam splitter must be determined. The relationship between the scatter-detector signal and sample scatter is found by misaligning the incident beam (of known power) so that it strikes the interior of the integration sphere. All three detectors must be linear over their range of use. Using this information, the relative values of the incident power P_i, the specularly reflected power P_0, and the scattered power P_s, may be found. When these signals are ratioed, values of the specular reflectance, diffuse reflectance, and total integrated scatter are obtained.

$$R_{spec} = P_0/P_i \qquad (6.15)$$

$$R_{diff} = P_s/P_i \qquad (6.16)$$

$$TIS = P_s/P_0 \qquad (6.17)$$

Advantages of this technique are ease of alignment (the scatter detector is relatively insensitive to position), reduction of the high-angle scatter problems mentioned in Sec. 1.6, and ready availability of diffuse integration spheres.

6.12 Error Analysis of the Measured BSDF

BSDF measurement variations as large as an order of magnitude have been reported (Leonard, 1988, 1989; Young, 1975) in round-robin tests, where several labs measured the same sample(s). Measurement variations have been so common that agreement within a factor of two is often viewed as *close enough for government work* (which is often the case). Wild variations in near-specular measurements have led to the (often correct) view that the lab with the lowest measured BSDF has the most accurate value. Because the calculation of BSDF is very straightforward, the source of these disagreements can be examined through a simple error analysis (Cady, 1989a).

$$\text{BSDF} = \frac{P_s/\Omega}{P_i \cos \theta_s} \tag{6.18}$$

$$\frac{\Delta\text{BSDF}}{\text{BSDF}} = \left[\left(\frac{\Delta P_s}{P_s}\right)^2 + \left(\frac{\Delta P_i}{P_i}\right)^2 + \left(\frac{\Delta\Omega}{\Omega}\right)^2 + \left(\frac{\Delta\theta_s \sin \theta_s}{\cos^2 \theta_s}\right)^2 \right]^{1/2} \tag{6.19}$$

Equation (6.18) has been found by standard error analysis (Squires, 1985) under the assumption that the four defining variables are independent of one another.

In similar fashion, the first term may be broken into the components that cause errors in the measured scatter signal. These are ΔP_y, the error caused by aperture misalignment in the y direction (perpendicular to the sweep direction), ΔV_s, the electrical noise generated in the receiver electronics, and NL, the fractional receiver nonlinearity.

$$\frac{\Delta P_s}{P_s} = \left[\left(\frac{\Delta I}{I}\right)^2 \left(\frac{\Delta P_y}{P_s}\right)^2 + \left(\frac{\Delta V_s}{V_s}\right)^2 + (NL)^2 \right]^{1/2} \tag{6.20}$$

The vertical misalignment term may be estimated everywhere (except the specular direction) by assuming that on a log-log plot the BSDF falls off as a straight line with slope M. It is not uncommon to find such straight-line segments with slopes varying from -1 to -3 (Sec. 4.5). Assuming circular symmetry in the scatter pattern in the near-specular region, the fractional error in the received power is

$$\frac{\Delta P_y}{P_s} = \frac{M}{\theta} \left[\cos^{-1}\left(\frac{R \cos \theta}{\sqrt{\Delta y^2 + R^2}}\right) - \theta \right] \tag{6.21}$$

where R is the receiver/sample distance and Δy is the vertical misalignment.

This source of error is small at large scatter angles, but can be significant near specular. The electronic noise term will dominate all BSDF measurements near the NEBSDF. However, since we are concerned with instrument errors where there is sufficient signal available, this term will be ignored. The nonlinearity term depends on the individual instrument. Most photovoltaic detectors and photomultiplier tubes are linear to within 1 or 2 percent for several decades until either saturation on the high-signal end, or the noise floor on the low-signal end are approached. The saturation point for each detector should be found experimentally to avoid the introduction of large errors.

The component for errors in P_i in Eq. (6.19) can also be broken into individual terms. If the *absolute* method is used to find P_i, these are

$$\frac{\Delta P_i}{P_i} = \left[\left(\frac{\Delta P}{P_i}\right)^2 + \left(\frac{\Delta V_s}{V_s}\right)^2 + (NL)^2 \right]^{1/2} \qquad (6.22)$$

The first term accounts for the loss of signal associated with the finite aperture of the receiver. For a sufficiently large aperture, this error can be kept well below 1 percent. The second and third terms are identical to those discussed above. If the reference sample method is used to measure P_i, one must loop back through the entire error analysis to evaluate this term. However, by looking ahead to the end of this calculation, errors of only a few percent are achievable under the optimum conditions that should be present for a reference sample measurement of P_i.

Uncertainties in the solid angle account for the third error component. These are caused by measurement errors in the receiver aperture radius r_s and in the aperture to sample distance R.

$$\frac{\Delta \Omega}{\Omega} = \left[\left(\frac{2\Delta r_s}{r_s}\right)^2 + \left(\frac{2\Delta R}{R}\right)^2 \right]^{1/2} \qquad (6.23)$$

The last component accounts for uncertainty in receiver position θ_s. Near the sample normal the error is quite small; however, it grows without limit near grazing scatter angles.

These errors are calculated in Table 6.2 at various values of θ for a typical situation described by constants $\theta_i = 5°$, $\Delta y = 25$ μm, $R = 50$ cm, $\Delta R = 0.1$ cm, and $\Delta\theta = 0.05°$. Values smaller than 1 percent are ignored. Very near specular the error exceeds 10 percent and is dominated by the contribution from aperture misalignment out of the incident plane. Near grazing angles the error is dominated by uncer-

TABLE 6.2   Example Error Values

$\theta = \theta_s - 5° =$	0.01°	0.1°	1°	10°	80°
M	2	2	2	2	2
Δr (mm)	5	5	20	50	50
r (mm)	150	150	800	5000	5000
$\dfrac{P_y}{P_s}$	0.15	0.002	0	0	0
$\Delta V_s/V_s$	0.001	0.01	0.001	0.001	0.001
NL	0.02	0.02	0.02	0.02	0.02
$\Delta P_T/P_i$	0.005	0.005	0.005	0.005	0.005
$\dfrac{2\Delta r}{r}$	0.067	0.067	0.050	0.020	0.020
$\dfrac{2\Delta R}{R}$	0.004	0.004	0.004	0.004	0.004
$\dfrac{\Delta\theta \sin \theta_s}{\cos^2 \theta_s}$	0	0	0	0	0.114
$\dfrac{\Delta\text{BSDF}}{\text{BSDF}}$	0.166	0.07	0.054	0.028	0.117

tainty in the value of θ_s and again exceeds 10 percent. However, between those two extremes the error stays well below 10 percent.

Combining Eq. (6.19) with Eq. (6.20) to (6.23) and eliminating the terms that give little contribution, result in the following expression for errors in the measured BSDF:

$$\frac{\Delta\text{BSDF}}{\text{BSDF}} \approx \left[\left(\frac{M}{\theta}\right)^2 \left[\cos^{-1}\left(\frac{R \cos \theta}{\sqrt{\Delta y^2 + R^2}}\right) - \theta \right]^2 + 2(NL)^2 \right.$$
$$\left. + \left[\frac{2\Delta r}{r}\right]^2 + \left(\frac{\Delta\theta_s \, \sin \theta_s}{\cos \theta_s}\right)^2 \right]^{1/2} \quad (6.24)$$

BSDF measurements are not inherently error prone. Over most of the angular range consistent results can (and have) been achieved. Detector saturation, ignoring convolution and signature effects, operator error resulting in gross increases to parameter uncertainties, software problems, lack of a common data format, and confusion about the definition of BSDF lead to most of the wide variations reported in the round-robin studies.

6.13 Summary

As with many other fields of metrology the computer has had an almost magical effect on the ability to produce fast, accurate scatter data in an expanding variety of forms. For the most part, modern instrumentation leaves the operator free to worry about sample-dependent issues. The key issues for operator concern are instrument calibration and instrument noise (or signature). Noise can be a significant problem when measurements are made very near specular. The techniques for overcoming signature and noise difficulties include the use of small apertures and focused beams, careful consideration of system geometry, the use of a mirror as a final focusing element, careful receiver design, low-noise electronics, and the use of system software that checks for signal deviation. These techniques have been outlined in the chapter. Calibration and checks on calibration have been discussed and an error analysis has been presented. Careful design will limit errors to about 5 percent except very near specular or at sample grazing scatter angles. In addition to the fairly common in-plane measurements, out-of-plane measurements, raster scans, retroreflection measurements, and integrating TIS systems have been discussed.

Additional measurement techniques are described in Chap. 8, which combines the more standard techniques presented here with some of the polarization relations given in Chap. 5. The next chapter covers scatter predictions based on scaling, curve fitting, and the known scatter behavior of optical components.

7

Scatter Predictions

One of the difficulties associated with scatter measurements is the large number of variables on which the scatter distribution depends. In addition to sample parameters, such as roughness, bulk defects, and contamination, there are instrumentation-dependent parameters, such as polarization, angle of incidence, and wavelength. Separating the various effects is not trivial, although there can be strong economic motivations to do so. The usual approach has been to make scatter measurements under the conditions expected for actual use—that is, to use the polarization, incident angle, and wavelength that are intended for eventual use to make the scatter measurements. Although polarization and incident angle are relatively easy to adjust in most instruments, a huge amount of data would be required to cover all of the combinations that a given sample might encounter in its expected use. In addition, generating scatter data at arbitrary wavelengths is an expensive task. So there are strong motivations for being able to predict sample scatter and avoid taking the data.

Unfortunately, there has been a tendency to oversimplify some scatter predictions, which has led to poor results and considerable confusion. For example, the relationship TIS = $(4\pi\sigma/\lambda)^2$ can be interpreted as follows: if the mean square roughness doubles, the measured TIS will also double. This is essentially true (within the limitations of TIS measurements given in Sec. 1.6) if the increase in σ is produced proportionally over the spatial bandwidth corresponding to the angular collection cone of the scatter instrument (i.e., the PSD doubles over the collection angles). But the equation is easily misinterpreted to indicate that if you double the illuminating wavelength the measured scatter from a given reflector should be reduced by one-fourth. Although one could probably guess that the TIS will be reduced in such a case, it would be just a guess. The equation is simply not appropriate for wavelength scaling, because it tacitly assumes a fixed spatial bandwidth. Equally misleading would be the assumption that scatter in a given direction will scale as $(1/\lambda)^4$ because of the leading factor in Eq. (3.43). Chaps. 1 to 4 indicate that the real issues are the shape of the PSD and which section of it scatters into the instrument at each wavelength. Because these concepts are not widely understood, wavelength scaling predictions have developed a much shadier reputation than their surface statistical cousins. In fact, as illustrated in Chap. 4, if surface statistics (from scatter) can be relied upon, then so can proper predictions based on changes in wavelength, angle of incidence, and polarization. The yellow brick road for these predictions is the Rayleigh-Rice BRDF/PSD relationship which is built on the bedrock of the smooth, clean, front-surface reflective (i.e., topographic scatter) conditions.

This chapter categorizes various aspects of making scatter predictions and gives guidelines for some cases. The Rayleigh-Rice predictions of the first section follow directly from the previous chapter. Other predictions are more speculative in nature, but not restricted to the well-behaved reflectors of Chap. 4. These include curve fitting of discrete data points, extension of BSDF data into unmeasured regions, and the use of sample characterization computer code (based on constants found from a few measurement scans) to predict any BSDF combination. Chapters 1 to 5 are required as background material.

7.1 Optical Surfaces: Using the Rayleigh-Rice Equation

If Secs. 4.1 to 4.4 have been carefully read, it will be clear that Eqs. (4.3), (4.4), and (4.11) can be used to predict an angle-limited section of the BRDF from a bandwidth-limited section of the sample PSD. The angle-limited blocks of scatter predicted in this manner will shift in location, width, and amplitude as a function of wavelength, incident angle, and polarization. It will also be clear that these predictions will be accurate only for the cases where the samples are smooth, clean, front-surface reflectors. Further, regardless of past prejudices, these predictions should be treated with the same degree of confidence that the scatter-to-surface statistics calculations are treated (Stover et al., 1984, 1988). If the surfaces are not smooth, clean, front-surface reflective (i.e., they do not scatter topographically) at all wavelengths, then other scaling laws must be found to predict their scatter (Church and Takacs, 1989; Stover et al., 1989). The predictions of this section are restricted to the cases where topographic scattering applies, although deviations from this special case will be noted.

Topographic scaling predictions are made directly from the PSD and the corresponding choices of wavelength, incident angle, and polarization (value or expression for Q). The result is an asymmetrical cone of scattered light about the specular reflection. Prediction of the scatter pattern includes values for the scatter direction (θ_s, ϕ_s) as well as the BRDF. The general, isotropic, and one-dimensional cases are treated separately.

7.1.1 The general case

Equation (4.3) is easily solved for the BRDF in terms of the PSD magnitude $S(f_x, f_y)$ and the two-dimensional grating equations [Eqs. (3.44) and (3.45)] can be solved for the corresponding scatter direction in terms of the PSD frequencies (f_x, f_y).

$$\text{BRDF} = \frac{16\pi^2}{10^8 \lambda^4} \cos \theta_i \cos \theta_s \, Q \, S(f_x, f_y) \tag{7.1}$$

$$\phi_s = \tan^{-1}\left[\frac{\lambda f_y}{\lambda f_x + \sin \theta_i}\right] \tag{7.2}$$

$$\theta_s = \sin^{-1}\left[\frac{\lambda f_x + \sin \theta_i}{\cos \phi_s}\right] \tag{7.3}$$

The 1, 2 subscripts of Chap. 2 on S have been dropped here. The dimension on S is inferred from the number of frequency components in the argument. In order to make a prediction, the PSD is inserted as a function, a curve fit, or as discrete points and the desired wavelength, angle of incidence, and polarization Q are used. If the surface is not isotropic, then each point in the predicted BRDF must be associated with a point in the measured BRDF. Wavelength changes affect first-order diffraction position and intensity. The scatter pattern will expand (or collapse) about the specular beam as the wavelength increases (or decreases). The pattern moves with the specular beam as θ_i changes, becoming less symmetrical as θ_i increases. Values of θ_s greater than 90 degrees correspond to light scattered into the surface and do not produce meaningful BRDF results. This is quite likely to occur if the values of θ_i or λ are increased from those used to obtain the PSD. In principal, scatter very near the specular beam can be predicted from scatter data taken at shorter wavelengths (Vernold, 1989). However, the usefulness of this technique is limited to some extent because the scatter very near specular is masked by the diffraction-limited specular focus, which expands as the wavelength increases.

There are some practical complications. In order to properly handle the variations in Q, the complex dielectric constant must be known as a function of wavelength and the values of Q evaluated by one of the techniques suggested in Sec. 5.2. If the PSD was obtained from discrete scatter data taken at evenly spaced points after the method of Chap. 4, the discrete PSD will be a series of unevenly spaced points. When these are used to evaluate a new BRDF, the points will again be unevenly spaced, which makes computer codes used to analyze and display the predictions more cumbersome.

7.1.2 The isotropic case

The equations of interest are identical to those used for the general case. The PSD is symmetrical and, except for normal incidence, the BRDF is not. Again, predictions are made by inserting the appropriate values of θ_i, λ, and Q. The quantity $S_{iso}(f)$ of Sec. 4.2, is not of interest because it is used only as a mathematical convenience for calculation of surface statistics.

Figure 7.1 shows plots of $S(f_x,0)$, a portion of the two-dimensional PSD calculated from BRDF data taken at two angles of incidence from the molybdenum mirror of Figs. 4.1, 4.5, and 4.6, and an aluminum mirror. The molybdenum mirror was shown to be linear-shift invariant in the visible and near-IR in Chap. 4. Because the PSDs for these mirrors are identical within the regions of overlap, it is reasonable to use them in Eq. (7.1) to predict scatter patterns associated with different angles of incidence. In Fig. 7.2 the molybdenum mirror is shown to exhibit wavelength scaling in the visible and near-IR. The same PSD is found at three different wavelengths. If the same PSD is found from each BRDF, it is clear that only one BRDF was needed to accurately

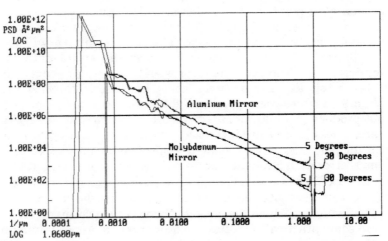

Figure 7.1 The PSD of two front-surface mirrors calculated from BRDF data taken at two angles of incidence.

COMPARE PSD

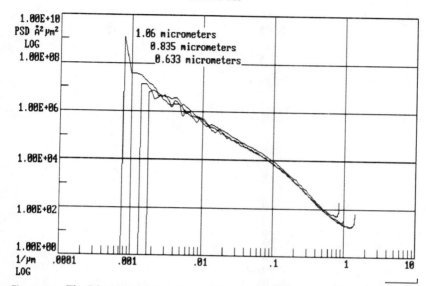

Figure 7.2 The PSD of the molybdenum mirror calculated from scatter data taken at three wavelengths in the visible and near IR.

predict scatter at any of the wavelengths. In these charts the PSD was evaluated after subtracting the instrument signature (multiplied by the sample reflectance) from the measured BRDF. Near specular a point is reached where the adjusted signature is equal to the BRDF. This point determines the low spatial frequency limit of each PSD curve and is determined by the various contributions to signature for each experimental setup, as well as the relationship between frequency and wavelength expressed in the grating equation. Also of interest is the fact that the PSDs of Figs. 7.1 and 7.2 are near-straight lines on log-log scales. They are fractal in nature with slopes of about -1.6 (see Sec. 4.5.2).

The fact that the molybdenum mirror scales at the three wavelengths shown in Fig. 7.2, does not mean that it scatters topographically at all wavelengths. As shown in Fig. 7.3, the same PSD is not found if the wavelength in increased into the mid-IR. It is difficult to say what causes the effect. It may be that the increased skin depth associated with the longer wavelength is enough to reach centers of subsurface scatter. Or variations in the optical constants across grain boundaries may be more significant in the mid-IR. The bottom line is that the molybdenum mirror scatters topographically in the visible and near-IR, but not in the mid-IR. The PSDs of the aluminum mirror (not shown) have similar variations when found from mid-IR scatter

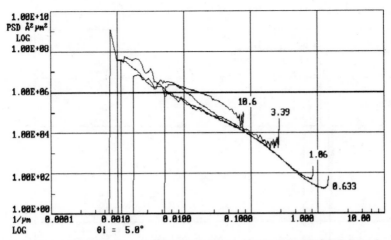

Figure 7.3 The PSD of a molybdenum mirror calculated from BRDF data obtained at four different wavelengths.

data. This type of behavior appears to be both sample and material dependent. No hard-and-fast rules have been established.

Figure 7.4 shows the measured BRDFs at 0.633 and 10.6 μm of a beryllium mirror plotted against the absolute value of $\beta - \beta_0$. The data were taken with an incident angle of 30 degrees from specular. The sample is linear shift invariant at both wavelengths. Figure 7.5 shows plots of the sample PSD calculated from the BRDF data of Fig. 7.4 and a second set of BRDF curves with an incident angle of 5 de-

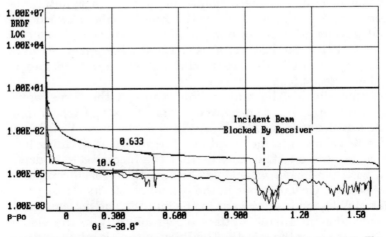

Figure 7.4 The BRDF of a beryllium mirror plotted against $1\beta - \beta_0 1$. The plus and minus sides are symmetrical.

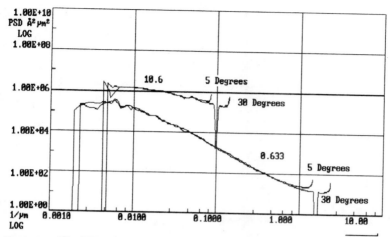

Figure 7.5 The PSD of a beryllium mirror formed from BRDFs taken at two different angles of incidence and two wavelengths.

grees. The PSDs are noticeably different for the wavelength change, but virtually identical for the incident angle change. It scales in incident angle, but not in wavelength. Although the sample does not scatter topographically (i.e., there are BRDF contributions that do not scale as fourth power of the inverse wavelength), the implication is that it is linear-shift invariant—or to put it differently, the grating equation(s) (and conservation of momentum) still apply. This means that we can expect linear-shift invariant properties to be present for samples that do not meet the smooth, clean requirements of the Rayleigh-Rice model. Figure 7.6 shows PSDs found from BRDF data taken at four different wavelengths for the beryllium mirror of Fig. 7.5. Remember that the PSD is a property of the sample—not the measurement technique. By definition the PSDs of Figs. 7.5 and 7.6 should be identical within their regions of overlap if the samples are actually smooth, clean, front-surface reflectors. The conclusion is that this beryllium sample does not meet these conditions and that probably none of the calculated PSDs in Figs. 7.5 and 7.6 are correct. If the trend observed for the molybdenum mirror is still in effect, the PSDs found from the short wavelength measurements are likely to be more accurate. The Rayleigh-Rice equation cannot be accurately employed for wavelength scaling from the data taken from this sample.

The nonconforming PSD curves are still of some interest. Notice that the calculated PSD grows with wavelength for both materials. This implies that there is extra scatter at the longer wavelengths. We know that topographic scatter falls off as the fourth power of the inverse wavelength. If nontopographic scatter is proportional to the in-

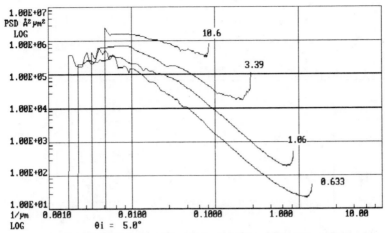

Figure 7.6   The PSD of the beryllium mirror of Fig. 7.5 taken at four wavelengths.

verse wavelength squared or cubed, then its contribution to the BRDF would become more evident at longer wavelengths—just as the data in Figs. 7.3, 7.5, and 7.6 indicate. Church and Takacs (1989) has postulated the following relationship for various weak scattering sources:

$$\text{BRDF} \sim \frac{1}{\lambda^n} S \frac{|\sin \theta_s - \sin \theta_i|}{\lambda} \tag{7.4}$$

The wavelength dependence of material properties is ignored in this equation. The value n is 4 for topographic and thin columnar defects, 3 for interference and random bulk defects, and 2 for thick columnar defects.

Because the molybdenum mirror scales topographically at several wavelengths in the visible and near-IR, the corresponding PSD should be correct. This means that in this wavelength region, scatter data can be used with Eq. (7.1) to find surface statistics or scatter in that wavelength band, but not in the mid-IR. And, using the techniques published by Church, it should also be possible to predict the scatter from surface profile data (Church, 1988, 1989). For samples like the beryllium mirror, the only safe method to obtain the sample BRDF is to measure it directly at the wavelength of interest, or to establish a new scaling law that incorporates a description of the appropriate nontopographic behavior. Unless it is already known that a sample is going to scale topographically, it is inappropriate to use surface profiling specifications to assure that a component will be low scatter or meet a given scatter specification.

7.1.3   One-dimensional samples

Equation (4.11) is solved for the BRDF in terms of the one-dimensional PSD magnitude $S(f_x)$ and the equation can be solved for the corresponding value of θ_s in terms of the PSD frequency f_x. The assumption has been made that the grating lines are fixed perpendicular to the incident plane.

$$\text{BRDF} = \frac{dP/d\theta_s}{P_i \cos \theta_s} = \frac{16\pi^2}{10^8 \lambda^3} \cos \theta_s \cos \theta_s \, Q \, S(f_x) \tag{7.5}$$

$$\theta_s = \sin^{-1}[\sin \theta_i + f_x \lambda] \tag{7.6}$$

A near-sinusoidal reflective grating can be used to demonstrate the validity of Rayleigh-Rice predictions for one-dimensional samples (Stover et al., 1988). The plus and minus first orders from a reflective aluminum grating were measured using incidence angles of 5 and 45 degrees and wavelengths of 0.488 and 0.633 μm. For the special case of a sinusoidal grating, Eq. (7.5) can be reduced to

$$P_{\pm 1}/P_i = \frac{(2\pi a)^2 \cos \theta_i \cos \theta_s \sqrt{R(\theta_i) \, R(\theta_s)}}{\lambda^2} \tag{7.7}$$

where an exact evaluation of Q (see Sec. 5.2) has been added to the reasoning behind Eq. (3.49). This form is convenient because it allows direct calculation of the grating amplitude a. Table 7.1 gives the results of these calculations. The average calculated value of grating amplitude is 211 Å with a standard deviation of about 1 percent. The agreement is excellent for both changes in wavelength and incident angle. Obviously the Rayleigh-Rice equation can be used in this situation.

TABLE 7.1   Calculation of Grating Amplitude

Order	θ_i	a @ 0.633 μm	a @ 0.488 μm
+ 1	5°	213 Å	211 Å
− 1	− 45°	210 Å	208 Å

7.2   Partial Data Sets

All BSDF measurements are incomplete in some way. Apparently continuous plots are usually the result of many closely spaced discrete measurements and there are always limits in scan length. In particular, near-specular measurements are limited by instrument signature and a finite width-receiver aperture (see Secs. 6.2, 6.3, and 6.4). There are situations where drastically reducing the number of data points, or extending the measured curve into difficult-to-measure regions, will offer large returns in time and cost with minimum risk.

7.2.1 Fractal surfaces

Fractal surfaces, which were discussed in Sec. 4.5.2, have the unique property that their PSDs follow power law relationships and can be expressed in terms of just two constants. A true fractal has a PSD that is a straight line on a log-log plot. The molybdenum mirror of Figs. 7.1 and 7.2 is a good example of a fractal-like surface. As pointed out in Sec. 4.5.2, there is an anomaly at zero frequency in the fractal relationship that will not be found in measured data. Power law behavior is not limited to just front-surface mirrors. It is not uncommon for transmissive samples to exhibit near-straight-line behavior of their BTDFs on log-log plots. The effect is common enough that scatter specifications (see Chap. 9) are often written in terms of a value and slope at a fixed angle of the measured BSDF.

It is particularly tempting to extend fractal behavior into the near-specular region where measurements are difficult to take. Section 9.2.3 gives an example where scatter specifications are developed using this technique. The point is that extending a constant slope PSD beyond its measured boundaries in order to predict additional sample behavior is not always as risky as predicting stock market trends. This is especially true if the material and manufacturing techniques have been known to previously produce the predicted PSD. Granted, sooner or later a *Black Monday* will be encountered by all speculators, but sometimes the odds are worth taking.

7.2.2 Curve fitting

Samples that have prominent spatial frequency components (on the surface or as bulk index fluctuations) will exhibit diffraction peaks that are difficult to predict from the measurement point of view (the precision-machined surface of Sec. 4.3, for example). Therefore, as recommended in Chap. 6, careful scatter measurement employs receiver step sizes that are smaller than the aperture width. This will typically result in as many as 500 to 1000 data points in a 90-degree scan. On the other hand, polished samples and even many one-dimensional samples, have BSDFs that are relatively smooth even though they

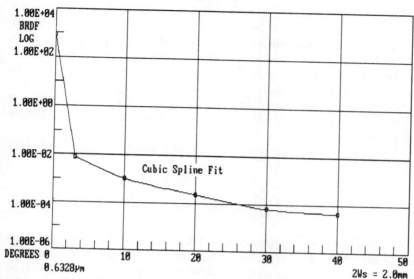

Figure 7.7   Curve fitting to BRDF of the molybdenum mirror.

vary over many orders of magnitude. Figure 7.7 shows the BRDF of the molybdenum mirror found by curve fitting five logarithmically spaced data points. The curve fit was done with a cubic spline. Essentially, this means that the fitted curve is forced to go through each data point with a slope equal to the straight-line slope obtained by connecting the two neighboring points. The slope at end points is the straight-line slope between the end point and its neighbor. The operation can be applied to the various plotting combinations (i.e., log-log, log-linear, etc.). Because many samples behave as fractals (linear PSD in log-log) the application of the curve fit is often more accurate when applied to the logs of the data. This also means that for a vast number of samples and process control applications only a few data points are necessary to get useful data sets or track changes in a production process.

7.3   Scatter from Nonspecular Samples

Diffuse samples scatter all of the specular light into the sphere about the sample. Samples may be reflective, transmissive, or translucent. In the case of a diffuse reflector the TIS is very large, because the power reflected in the specular direction is very small. Thus diffuse surfaces are by definition optically rough. The reflectance or transmittance of optically smooth samples can be accurately estimated by ratioing the specularly reflected (or transmitted) power to the incident power. Diffuse samples require the entire hemisphere (or sphere) to be considered in order to measure the total reflectance or transmittance. In the case of translucent materials, secondary scattering may cause the scattering area (or volume) to be larger than the area (or volume) illuminated by the incident beam (see Fig. 7.8). The combination of these effects makes the accurate prediction of scatter from diffuse (rough) components very difficult.

The hemispherical reflectance R from an opaque diffuser can be defined as

$$R = \frac{1}{P_i} \int\limits_{0}^{\pi/2} \int\limits_{0}^{2\pi} (dP_s/d\Omega_s) \sin \theta_s \, d\phi_s d\theta_s. \qquad (7.8)$$

in terms of the spherical geometry defined in Figs. 1.6 or 6.12a. This contains all of the reflected light—specular and scattered. The incident power, the reflected specular power, and the diffusely reflected power may be combined into ratios to form the diffuse reflectance, the specular reflectance, and the TIS. Similar definitions may be made for diffusely transmitting samples. In general, the resulting scatter pat-

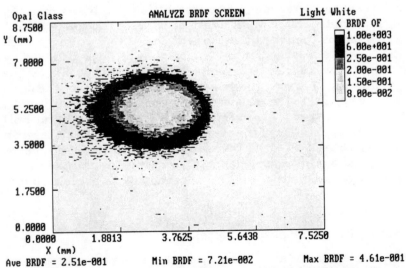

Figure 7.8 A piece of translucent opal glass has been raster scanned using the CCD technique of Sec. 6.9. The bright ring around the central disk is an artifact of the BRDF calculation. Multiple bulk scattering causes light to be radiated from outer regions where little or no light is incident. Thus, with P_i near zero, a large BRDF is calculated.

terns can be very complex. Fortunately, many diffuse materials do not scatter in arbitrary fashion. The easiest assumption to make about a diffuser is discussed in the next section.

7.3.1 Lambertian samples

A common assumption that is fairly reasonable for many diffuse samples is to define the scattered radiance to be a constant. This means that the scattered power/unit solid angle falls off as $\cos \theta_s$. In other words, as an observer of the illuminated spot moves in increasing θ_s, toward the waist of the hemisphere, the measured light intensity falls off in proportion to the apparent size of the radiating source. At $\theta_s = 90°$, the source is viewed on edge (zero apparent area) and the scatter signal drops to nothing. This relationship is assumed true regardless of the source incident angle. Samples which scatter in this manner are known as Lambertian samples. If the scattered intensity is proportional to $\cos \theta_s$ then the BRDF is a constant.

$$\text{BRDF} = F = \frac{dP_s/d\Omega_s}{P_i \cos \theta_s} = \text{constant} \qquad (7.9)$$

The value of the BRDF may be evaluated in terms of the hemispherical reflectance by substituting $F \cos \theta_s$ into Eq. (7.8) for the normalized scattered intensity and solving for F.

$$R = \int_0^{\pi/2} \int_0^{2\pi} (F \cos \theta_s) \sin \theta_s \, d\phi_s d\theta_s \qquad (7.10)$$

$$F = R/\pi \qquad (7.11)$$

Further, if the sample has completely diffused the incident beam the scattered polarization will be independent of the incident polarization. Using subscripts on F to denote the incident and scattered polarizations, as we did with Q, gives

$$F_{ss} = F_{sp} = F_{ps} = F_{pp} = R/(2\pi) \qquad (7.12)$$

Thus a good white diffuser will not have a BRDF exceeding $1/\pi$ and a low-reflectance diffuser (say minimum 1 percent) will not have a BRDF less than about $3 \times 10^{-3} \, sr^{-1}$.

Figure 7.9 shows the measured BRDF of several white diffusers on linear scales. The BRDF is reasonably flat over most of the hemisphere. The hemispherical reflectance may be estimated by using Eq. (7.11) or evaluated by integrating the curves as though the measurement is a sample on the (assumed) symmetrical scatter hemisphere. Figure 7.10 shows the BRDF of several black diffusers. The rise in

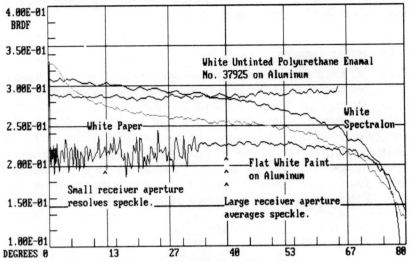

Figure 7.9   The BRDF of several diffuse white reflectors.

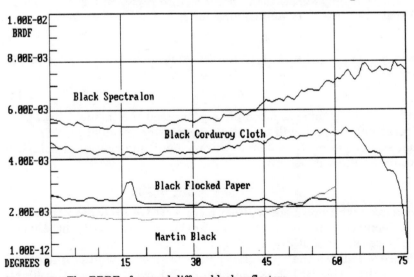

Figure 7.10   The BRDF of several diffuse black reflectors.

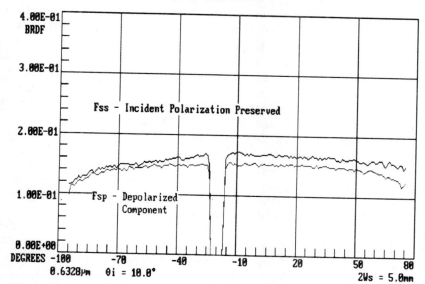

COMPARE SCATTER DATA

Figure 7.11 Depolarization of scatter light from a white spectralon sample is shown by rotating an analyzer in the receiver to accept *s* or *p* scattered light. The source was *s* polarized.

BRDF at high angles is common among black diffusers. Figure 7.11 shows the depolarization of the scattered light from a piece of white paper.

Using the above relationships and a book value (or estimate) of the hemispherical reflectance allows rough predictions of diffuse scatter to be made. Simply knowing the maximum/minimum bounds (which are less than three orders apart) may be enough to determine whether or not measurements are required. Many diffuse samples are not Lambertian enough to make use of this easy assumption. Another approach is required for these situations.

7.3.2 Non-Lambertian samples and material signatures

Many infrared and visible remote sensing applications require a rather complete knowledge of the BRDF of many different materials. For example, military camouflage netting and paint does its job over the visible wavelengths, but are those objects invisible to a satellite using an infrared sensor? What is the optimum wavelength to use to check for the spread of spotted knapweed in the western United States? To answer these types of questions, BRDF data are needed for essentially all incident directions and polarizations, all scatter directions and polarizations, and over many different wavelength bands for a huge variety of materials. A complete data set for a given material is known as the *material signature*. Collection, and even storage, of such a data set for even one material is obviously impractical. Instead, a model is developed that allows the calculation of the BRDF at the particular directions, polarizations, and wavelengths of interest. Based on the last section, it is easy to envision a straightforward model based on the Lambertian assumption. All that would be needed as input to the model is the hemispherical reflectance over the required wavelength regions. Unfortunately many of the materials in question fall into that broad region between optically smooth and ideally diffuse, where an assumption of Lambertian behavior would be inappropriate over at least some of the required wavelengths.

The obvious approach is to use an essentially empirical, computer model that requires a relatively small amount of measured data to predict the desired BRDF values. A few carefully chosen data scans are taken as input to the model. Model parameters are calculated from these data and the result is used to predict the BRDF. The model is easily checked for a few situations by comparing results to additional experimental data. Taking even the reduced set of data required for the model is not an easy task. Measurements must be taken out of the incident plane and at several wavelengths. A review of some of the experimental issues is found in Sec. 6.8.

Material signature codes are obviously of great interest to the government (DOD, NASA, etc.) and also to many of their contractors. The manufacturers of military aircraft, for example, consider that this type of information offers a strong potential competitive edge. As a result, although several material signature codes exist, very few of the algorithms on which they are based have been published. Two exceptions are the Maxwell-Beard model developed at the Environmental Research Institute of Michigan (Maxwell et al., 1973) and the code developed at McDonnell Douglas by Leader (1979).

7.4 Software for Prediction of Scatter in Optical Systems

Analysis of scatter in optical systems is reasonably straightforward if only two or three components are involved. The specification examples in Chap. 9 illustrate back-of-the-envelope calculations that can be used for simple systems. However, as the number of system elements increases, analysis by hand quickly becomes unmanageable. Those familiar with the use of ray-tracing software will appreciate the use of a computer to handle this type of calculation-intensive problem.

Scatter prediction and analysis are more complicated than ray tracing. Ray-tracing programs simply follow the laws of reflection and refraction to determine ray location (which is more difficult than it sounds when hundreds of rays and tens of components are involved). Scatter analysis requires that all or part of each ray be reduced to a scatter pattern at each component. The scattered light and the original specular ray must be accounted for as they propagate through the system. Scatter-prediction codes make use of various statistical techniques. Component BSDF data (or predictions) and the usual ray-tracing information (component location, radius of curvature, etc.) are used to define the system. The number of rays and directions are often limited (at the operator's discretion) to examination of scatter from just a few *critical objects*. Using the programs can be a little like playing a musical instrument, in that there is a fair amount of operator influence on the obtained output.

Stray-light analysis codes were developed in the early 1970s to improve baffles in space telescopes (Breault, 1986). Since then they have proved useful in a variety of optical design problems and play an important role in the early design stages of sophisticated optical systems. Several codes are now commercially available for use on VAX- and PC-compatible computers. Table 7.2 summarizes some of the better-known codes.

TABLE 7.2   Summary of Several Scatter-Analysis Codes

Name	Developer	Comments/references
APART	University of Arizona	(Breault et al., 1986)
APART/PADE	BRO Inc.	(Breault et al., 1986)
ASAP	BRO Inc.	(Greynolds, 1980)
FOOPAC	University of Arizona	(Foo, 1985)
GUERAP III	Honeywell/Telic	VAX/PC (Likeness, 1977; Freniere, 1980)
MINI-APART	Ball Corp.	(Bamberg, 1983)
ORADAS	Hughes	(Rock, 1986)
OSAC	Perkin Elmer	(Noll, 1982)
SOAR	Telic Optics	PC-based (no references before 1990)
STRAY	NASA Ames	FORTRAN 77, VAX11/780 (St. Clair Dinger, 1986)
TRAZ	El-Op	(Dolan, 1986)

7.5 Summary

The precision with which BSDF data can be predicted depends on the accuracy with which the sample can be modeled. Near-specular samples which fit the Rayleigh-Rice topographic criteria (smooth, clean, reflective) are the easiest to predict. In fact, confirmed predictions on samples that are known to fit the criteria, constitute a good check on instrument calibration. The general case for this model is conveniently broken into isotropic (polished) and one-dimensional (gratinglike) subsets that cover many practical applications. Fortunately, many near-specular samples are fractal in composition. In effect, defect amplitude (which is responsible for scatter amplitude) is related to defect width (which is responsible for scatter direction) by a geometric relationship. Because these relationships exist in many places throughout nature, the scatter of many samples can sometimes be predicted over wide ranges from a small section of data. Rougher (diffuse) samples are subdivided into Lambertian and non-Lambertian classes. Lambertian samples scatter at constant radiance regardless of angle of incidence. Scatter from near ideal Lambertian surfaces of known hemispherical reflectance is very easy to predict. Just the reverse is true for samples that are neither specular nor perfectly diffuse. Characterizing this huge class of samples has become increasingly important for remote sensing applications. The only practical way to obtain the full BRDF characteristic (or material signature) of such samples is by means of computer codes that use a limited set of BRDF data to predict the complete material signature. The software and algorithm problems are not trivial, and as was seen in the last chapter, neither is the hardware required to supply the multiple wavelength, variable polarization, and out-of-plane input data. Scatter codes now exist that make use of all of the above observations to predict scatter in optical systems. The codes combine the known BSDF, or predicted BSDF, of all system elements with ray-tracing techniques to provide designers with system limits prior to reduction to hardware. Originally written to develop telescope baffles, these programs are now being applied to a wide variety of optical-system scatter problems.

Detection
of Discrete Surface
and Subsurface Defects

Chapters 2, 3, and 4 concentrated on the relationship between scatter and smooth-surface topography. However, another extremely useful application of light-scatter metrology is the detection and mapping of component defects that do not meet the smooth, clean, reflective conditions of mirror surfaces. Examples of such defects are surface contaminants, scratches, digs, coating globs, and residues. In these measurement situations, scatter from the surface topography is considered background noise and the defect scatter is signal. Although defects often scatter more light than the surrounding surface topography, they may sometimes scatter considerably less light because they have a small cross-sectional area or because they are buried just beneath a reflective surface. In such cases, a low signal-to-noise ratio results. If it can be established that nontopographic defects scatter light differently than surface topography, these differences can be exploited to improve signal-to-noise and map the defects using the raster techniques described in Sec. 6.9. This chapter discusses the differences in topographic and defect scatter and outlines techniques that have been used to enhance defect detection.

One way that has been used to improve discrete defect signal to noise is to cross polarize the source and receiver. This technique has been employed successfully for a variety of applications. It has been used to separate the specular and diffuse return of radar signals from the moon (Mathis, 1963) to infer the relative amounts of moon rock and dust. The cross-polarization technique is used as a standard scan to check for Lambertian scatter and subsurface scatter, as part of the Maxwell-Beard model for obtaining material signatures (see Sec.

7.3.2; Maxwell et al., 1973). In terms of the vector/matrix polarization approach of Sec. 5.3, the cross-polarization technique amounts to evaluation of element s_{zz}. The method works, because for the correct choices of incident polarization and observation angle the scatter from surface topography can be virtually eliminated while enough cross-polarized defect scatter remains to dramatically increase signal to noise. When combined with raster, or fast raster measurements (Sec. 6.9), the result is a sensitive measure of defect size and location. Results can be analyzed further to produce statistics describing sample defects (size, density, etc.).

A general understanding of the material in Chaps. 1, 3, 4, 5, and 6 is required for this chapter. The next section gives an arm-waving explanation of the differences in defect and topographic scatter and why the cross-polarization technique works. Later sections give results obtained for large surface defects (height $\gg$ wavelength), particulate contamination, subsurface defects in transparent optics, and subsurface defects in opaque materials.

8.1   Polarization Effects Associated with Defect Scatter

As mentioned above, the cross-polarization technique has been employed by several different groups since at least the 1950s. For the technique to work well, some care has to be given to the choice of measurement parameters such as receiver position, incident angle, and aperture size and shape. As might be expected, the optimum choice of these parameters is sample dependent. However, some general guidelines can be formulated by analyzing the material presented in the preceding chapters and that is the subject of this section.

In Chap. 3, we saw that scatter from smooth surfaces (diffraction) occurs because of the phase changes introduced into the reflected light by the small deviations in path length caused by surface topography. Scatter from other sources can also be explained by induced phase and amplitude changes. For example, even a perfectly smooth surface that had random, and possibly abrupt, changes in dielectric constant would scatter light. Grain boundaries, which are evident on the surface of many metallic mirrors, can provide a close approximation of this situation. As illustrated in Fig. 8.1, scatter can be caused by many different types of variation in material surface and bulk. If these variations are mild, they can be characterized by weak single-scatter events, as in the case of smooth-surface topography, and polarization changes may not occur (Church and Takacs, 1989). On the other hand, if the defects are more pronounced, they may be characterized by multiple scatter events and dramatic polarization changes can be de-

Scatter Sources

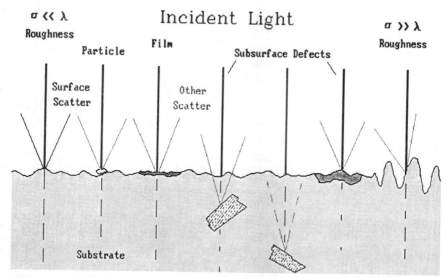

Figure 8.1 Scatter is caused by a variety of surface and subsurface imperfections. Only the smooth, clean, front-surface topography has been used to predict the corresponding scatter. Scatter from other defects can be fully characterized, but is more difficult to relate to defect parameters.

tected. The following discussion concentrates on polarization differences between smooth-surface (topographic) scatter (which has been extensively analyzed) and stronger sources of scatter (particulates, subsurface damage sites, etc.) that can be exploited to enhance defect detection. The object is to identify locations on the scatter hemisphere in front of a smooth, clean front-surface mirror that, for certain source conditions, will contain only one linearly polarized component. This component can then be rejected with an analyzer in front of the receiver. Defects which do not scatter to this location with the same polarization state can then be observed without interference from surface scatter.

The smooth-surface requirement virtually assures that all surface scatter is the result of single reflections from the surface. For all but the very highest scatter angles, there is no chance for reflected light to encounter the surface a second time. The result, as described by Eqs. (5.14) to (5.17) for the polarization constant Q, is that incident s-polarized light diffracted into the plane of incidence is still s polarized. The polarization vector of incident p-polarized light scattered into the incident plane is also unaffected. Light scattered out of the plane from

either of these incident beams will contain a cross-polarized component. This is expressed in Eqs. (5.15) and (5.16) by Q_{sp} and Q_{ps} taking
on zero values in the incident plane and nonzero values for out-of-
plane scatter. Linearly polarized light, that is incident with both s and
p components, is elliptically polarized upon reflection because the relative phase between the two vectors δ changes. At least three types of
defects can be readily identified that do not scatter in this manner.

If the surface features are rough, light may be reflected (scattered)
several times before leaving the surface. For example, consider the following scenario. An incident s-polarized ray of light enters a relatively deep-surface feature and is reflected off a wall in an out-of-
plane direction. It now contains both s and p components. It strikes
the far wall of the deep-surface valley and is reflected away from the
sample and out into a plane that is parallel to, and just slightly offset
from, the sample plane of incidence. This second reflection further
changes the polarization vector of the ray. The (now) elliptically polarized ray is then measured as plane-of-incidence scatter that contains both s and p components. Multiple scatter events involving
many surface reflections are responsible for depolarizing the light
scattered by diffuse surfaces and explain why the ideal Lambertian
surface scatters light equally in p and s components regardless of the
source polarization state. Rough surfaces, and rough (deep or high) defects will not preserve the incident s- or p-polarized state upon scattering into the incident plane.

The subject of scattering by small particles is an area of ongoing research and even a cursory explanation is outside the scope of this text.
Bohren and Huffman (1983) and van de Hulst (1957) have published
texts on this subject. For our purposes, it is enough to say that for the
general case of arbitrarily shaped particles larger than a wavelength,
incident s and p polarization are not preserved in the scatter pattern
even in the incident plane. The effect is easily observed by placing
small particles between two polarizers and holding them up to a light.
When the polarizers are rotated to the crossed position, the particulates are seen as bright dots on a dark field. Thus, just as for rough
surfaces, the *noise* associated with surface topography can be separated from a portion of the *scatter signal* associated with a surface particulate.

Light that is transmitted into a material is reduced exponentially in
distance by absorption as described in App. A. The skin depth is the
distance required for an intensity reduction of e^{-1}. In the visible the
skin depth varies from tens of angstroms in metals to a few thousand
angstroms in semiconductors, to as much as a few meters in transparent dielectrics. Subsurface defects in a uniform substrate will scatter
like particulates. Some of the light that is scattered from the defect

back toward the surface is transmitted out of the substrate material and can be detected to show defect location. This light amounts to back scatter from a particle (adjusted by Snell's law). As indicated above, the incident polarization will not be preserved in the plane of incidence. Thus, even in opaque materials, defects beneath the surface can be detected if they are within roughly a skin depth of the surface. Elimination of surface scatter in these measurements is particularly important because the scatter levels from subsurface defects can be relatively low.

The key to improving defect signal to noise is to remove as much of the surface topography scatter as possible. The equations for the polarization constant of Chap. 5 suggest more than one way to do this as a variety of observation, and source polarizations and directions can be used. The technique may be illustrated by considering the situation where s-polarized light is incident on a surface with several discrete defects, as shown in Fig. 8.2. A scatter receiver is centered on the surface normal with a slit aperture in the plane of incidence. Light entering the aperture is transmitted through a receiver analyzer that is oriented to pass only horizontally polarized light. If there are no defects present, most of the aperture light will be s polarized and none of it will be transmitted to the detector. For two reasons, a small amount of surface scatter will be transmitted to the receiver. First, the crossed polarizers (source and receiver) do not give zero extinction of the s-polarized light. And second, because the slit aperture must have a finite width, a small amount of out-of-plane light, with some p-polarized

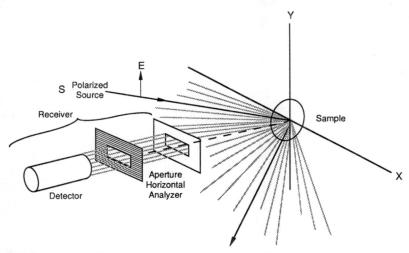

Figure 8.2 An s-polarized (vertical) source and a p-(horizontal) analyzed receiver in the plane of incidence. The aperture is shown as a narrow rectangle centered in the plane of incidence at the surface normal.

topographic scatter, will be passed to the detector. Because the defect-dependent fraction of incident-plane scattered light that is converted from s to p is unknown, and is often relatively small, it is worth looking more closely at the effect of these two noise sources.

In order to analyze the situation, the receiver signal must be obtained by integrating the scattered light over the receiver aperture and then passing it through the analyzer. An expression for the signal-to-noise ratio can be derived in terms of sample and system components. Of particular interest is the optimum out-of-plane opening of the aperture. Too small an opening will drop the signal below the electronic noise floor and too large an opening increases unwanted surface scatter. Figure 8.3 illustrates the situation of Fig. 8.2 with a bow-tie-shaped aperture centered on the surface normal. This shape is easier to analyze than the rectangle of Fig. 8.2. The angular extent of the aperture is $2\Delta\phi_a$ and $2\Delta\theta$. F' is used to denote the cosine-corrected BRDF, with subscripts D and T used to indicate defect and topographic scatter, respectively. The polarization components are indicated as before by s and p. The extinction ratio of the two polarizers is given by ξ and is equal to the ratio of minimum-to-maximum transmission of a crossed polarizer pair as one is rotated against the other. The light-power equivalent value of the background electronic and detector noise is denoted by P_{NE}. The signal-to-noise ratio can then be expressed as

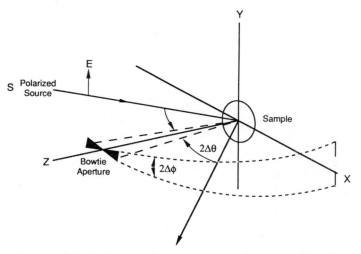

Figure 8.3 The shaded bow-tie aperture has dimensions of $2\Delta\theta$ and $2\Delta\phi$. This aperture shape is easier to analyze than the rectangle of Fig. 8.2 and restricts all values of ϕ to less than $\pm\Delta\phi$.

$$S/N = \frac{4P_i \int\limits_0^{\Delta\theta} \int\limits_0^{\Delta\phi} (F'_{Dsp} + \xi F'_{Dss}) d\Omega_s}{P_{NE} + 4P_i \int\limits_0^{\Delta\theta} \int\limits_0^{\Delta\phi} (F'_{Tsp} + \xi F'_{Tss}) d\Omega_s} \qquad (8.1)$$

The extinction ratio is applied only to the ss components. The various terms can then be evaluated with the use of several simplifying assumptions. The term $\xi F'_{Dss}$ will be dropped because it will almost always be significantly smaller than F'_{Dsp} for a typical value of ξ. Although the value F'_{Dsp} is unknown, it is reasonable to assume that it is constant over the aperture and can be brought outside the integral. This may not be true if the aperture is too close to the reflected specular beam. The values of F'_{Tss} and F'_{Tsp} can be evaluated from Eq. (4.1) giving surface scatter in terms of the polarization factor Q and the power spectrum S.

$$F' = F \cos\theta_s = \frac{16\pi^2}{\lambda^4}\cos\theta_i \cos^2\theta_s \, QS \,(f_x, f_y) \qquad (8.2)$$

To ease the calculation, the sample is assumed to be a good conductor and thus have polarization constants that are given by the simplified expressions of Eqs. (5.22) to (5.25).

$$Q = Q_{ss} + Q_{sp} = \cos^2\phi_s + \left(\frac{\sin\phi_s}{\cos\theta_s}\right)^2 \qquad (8.3)$$

It is further assumed that $S(f_x, f_y)$ is a constant over the limited bandwidth represented by the aperture. Thus the signal-to-noise ratio becomes

$$S/N =$$

$$\frac{4P_i F'_{Dsp} \int\limits_0^{\Delta\theta} \int\limits_0^{\Delta\phi} \sin\theta_s \, d\phi_s d\theta_s}{P_{NE} + 4P_i \left(\frac{16\pi^2 \cos\theta_i \, S}{\lambda^4}\right) \int\limits_0^{\Delta\theta} \int\limits_0^{\Delta\phi} [\xi \cos^2\theta_s \cos^2\phi_s + \sin^2\phi_s] \sin\theta_s \, d\phi_s d\theta_s}$$

$$\qquad (8.4)$$

where the differential aperture is

$$d\Omega_s = \sin\theta_s \, d\phi_s d\theta_s \qquad (8.5)$$

The bow-tie-shaped aperture eases integration because the aperture boundaries (integration limits) are along lines of constant θ_s or ϕ_s on the scatter hemisphere. The shape of the aperture fixes $\Delta\phi_s \ll \Delta\theta_s$ so that the small-angle approximation will be used for ϕ_s but not for θ_s. Substituting $\cos \phi_s = 1$ and $\sin \phi_s = \phi_s$, and then integrating gives

$$S/N = \frac{4P_i F'_{Dsp} \Delta\phi[1 - \cos (\Delta\theta)]}{P_{NE} + \frac{4}{3} P_i \left(\frac{16\pi^2 \cos \theta_i S}{\lambda^4}\right) \{\xi\Delta\phi[1 - \cos^3 (\Delta\theta)] + \Delta\phi^3[1 - \cos (\Delta\theta)]\}}$$

(8.6)

This equation may be interpreted as follows. The maximum value of $2\Delta\theta$ is limited by the width of the receiver. As expected, the signal (numerator) increases as the aperture is opened in either ϕ_s or θ_s. The second and third denominator terms (in square brackets) are noise due to ss surface light transmitting through the analyzer and sp surface light coming through the aperture just out of plane. As $\Delta\phi$ is increased, the third term grows faster than the second. The signal-to-noise ratio will grow as $\Delta\phi$ increases, until either the second or third denominator terms are approximately the size of P_{NE}. When the noise is dominated by the second term, the signal to noise is roughly constant. Then further increases in $\Delta\phi$ will eventually result in the noise being dominated by the third term, the signal to noise will decrease, and smaller defects will start to be lost in surface scatter. So not only is the bow-tie shape easier to analyze, but it also reduces noise. It is generally useful to open the aperture in $\Delta\phi$ to the point where the second and third terms are about equal. This occurs at

$$\Delta\phi = \left[\xi\left(\frac{1 - \cos (\Delta\theta)}{1 - \cos^3 (\Delta\theta)}\right)\right]^{1/2} \simeq 3° \text{ for } \cos (\Delta\theta) = 0.9 \text{ and } \xi = 10^{-3} \quad (8.7)$$

The relatively small value of $\Delta\phi$ justifies the earlier small-angle approximation. Crossed polarizers with extinction ratios as low as $1/10^5$ are available in the visible. In the infrared, ratios of about $1/10^2$ are more common, but units exceeding $1/10^4$ can be purchased. The signal to noise can be peaked by setting the differential of Eq. (8.6) with respect to $\Delta\phi$ equal to zero and solving for $\Delta\phi$.

$$\Delta\phi_{\text{peak}} = \left[\frac{3\pi^2 P_{NE}}{k^4 \cos \theta_i S}\right]^{1/3} = \left[\frac{P_{NE}}{F_{ss}}\right]^{1/3} \quad (8.8)$$

The values P_{NE} and S (or F_{ss}) can be found experimentally. The larger of the two values of $\Delta\phi$ found from Eqs. (8.7) or (8.8) should be used to obtain maximum signal at a good signal to noise.

Obviously other combinations exist. A bow tie oriented vertically ($\phi_s = 90°$) for the input situation of Fig. 8.2 gives identical results (remember that the directions defining s and p change with ϕ_s). Values of

ϕ_s between 0 and 90 degrees result in excessive surface scatter getting through the analyzer. The situation for a p-polarized input is more complicated because of the addition of terms required for Brewster's angle; however, similar results are obtained. p-Polarized light incident at Brewster's angle will reduce the near-specular surface reflection, but Q_{pp} away from the incident angle is not zero so *surface noise* is still present in other directions. Other zero-noise locations can be found elsewhere on the scatter hemisphere.

8.2 Bulk Defects in Transparent Optics

For many applications involving transmissive optics, it is useful to separate scatter caused by surface roughness from that due to bulk defects. For example, substrates to be coated for use as low-scatter reflectors (such as ring-laser gyroscope mirrors) are more sensitive to surface defects than bulk defects. Damage just below the surface of a polished optic is often caused in fabrication. Changes in fabrication technique intended to reduce the generation of subsurface defects are difficult to monitor without the ability to separate surface and subsurface scatter.

Figure 8.4 shows the top view of a laser beam passing through a transparent sample. The incident laser beam strikes the front surface, propagates through the bulk, and exits through the back surface. Multiple reflections are not shown in this figure. The sample scatters

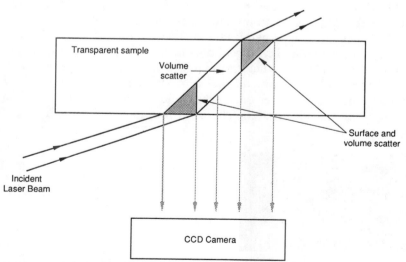

Figure 8.4 Top view of the laser beam passing through a transmissive sample showing regions of bulk scatter and regions of surface *and* bulk scatter.

enough light so that a CCD camera and an 8-bit frame grabber can record the image and send the information to a computer. A combination of bulk and surface scatter are viewed at the ends of the illuminated volume (shaded areas) and bulk scatter alone makes up the center section. It is apparent from this figure that significantly large amounts of scatter from the bulk could totally obscure measurement of scatter from the surface. One method used to calculate the surface scatter is that of subtracting bulk scatter measured in the center of the image from a surface and bulk-scatter combination measured at the ends (Orazio et al., 1982). Unfortunately, this method sometimes results in negative values for the calculated surface scatter. The negative values arise from uncertainties in geometry, beam profile, and from problems associated with subtracting noisy signals of similar magnitude.

The measurement can also be made by applying a variation of the cross-polarization technique (McGary et al., 1988). Using an s-polarized source, a receiver analyzer, and an aperture that admits only light in or near the incident plane, three images of the sample are recorded. The first is with the receiver analyzer removed. The second is with the analyzer in place and oriented to pass s-polarized light. The third is with the analyzer rotated to pass only the cross-polarized p component. The center of the beam, as recorded in the two cross-polarized measurements, is used to determine the relative intensity of s- and p-polarized light from the bulk scatter. This ratio is then multiplied times the third image to obtain the total bulk scatter (s and p) without any surface scatter present. This image is then subtracted from the first (total scatter) image. The difference is an image of the surface scatter alone. Additional adjustments must be made to the images to compensate for transmission of the analyzer, changes in camera integration time, and changes in camera noise level with integration time. Figure 8.5 is a photograph of the video monitor screen with the resulting total, bulk, and surface scatter images superimposed for a Zerodur flat. Horizontal cross sections taken from the data shown in Fig. 8.5 are plotted in Fig. 8.6. The solid line in Fig. 8.6a is a plot of

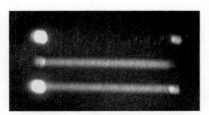

Figure 8.5 Total scatter (top), volume scatter (middle), and surface scatter (bottom) displayed on a video monitor.

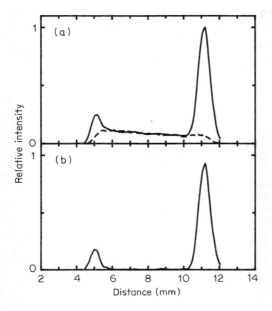

Figure 8.6 Plots of horizontal cross sections taken from data shown in Fig. 8.5. (*a*) The total scatter (solid line) and the volume scatter (dashed line) are plotted; (*b*) the surface scatter is plotted. Scatter from the front surface is seen on the left in (*b*), and scatter from the rear surface is seen on the right.

the total scatter while the dashed line shows the bulk scatter. The surface scatter is shown in Fig. 8.6*b*. The front surface shown on the left side has a better polish than the back surface and scatters less. The high scatter signal from the back surface is visible on the right.

The transmitted beam is reduced in power as it passes through the volume, due to a combination of scattering and absorption. The data also provide a means for calculating the exponential loss coefficient α for the material. Assuming the bulk to be homogeneous and isotropic, light propagating through the bulk will decrease exponentially in intensity due to absorption and scatter (see App. A). The loss coefficient is obtained from the image of bulk scatter. Figure 8.7 is a plot of beam intensity as a function of distance in the Zerodur sample and was taken from Fig. 8.5. An exponential was fit to the center portion of the curve and the loss coefficient found to be 0.0178 mm^{-1}. This coefficient can also be calculated using the measured values of sample reflectance, transmittance, and thickness, which for the Zerodur sample gives a value of 0.0181 mm^{-1}, confirming the previous technique.

Similar profiles of the volume and surface scatter from a zinc selenide window are shown in Fig. 8.8. The volume-scatter profile peaks just under both surfaces. This is probably due to subsurface damage that occurred during polishing. Polycrystalline substrates are particularly sensitive to this kind of damage. There is no evidence of large discreet defects in this first sample, but the zinc selenide of Fig.

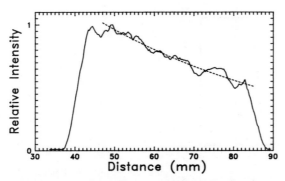

Figure 8.7 A horizontal cross-sectional plot of the bulk-scatter profile is given by the solid line. Data were fit with an exponential given by the dashed line to obtain the extinction coefficient, 0.0178 mm^{-1} for the Zerodur sample.

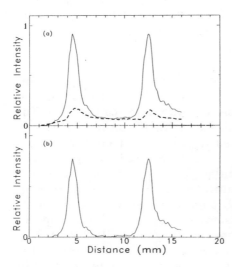

Figure 8.8 Plots of total, volume, and surface scatter for a zinc selenide window. In (a), the total scatter (solid line) and volume scatter (dashed line) are plotted, and in (b) the surface scatter is plotted. The front and back surface scatter are approximately the same because of uniform finishing. The volume scatter peaks just under both surfaces.

8.9 shows several bulk defects. These two windows were part of a set of samples that had been polished to similar surface finishes, but had different bulk qualities. Figure 8.10 shows BTDF scans of these samples taken at 0.633 μm. Each sample was scanned twice, once to get F_{ss} and once to get F_{sp}. Notice that for these samples the ratio of F_{ss} to F_{sp} is about the same (remember the log scale). This indicates that the bulk defects are relatively weak in nature. They do not cause multiple scatter events that are responsible for depolarization.

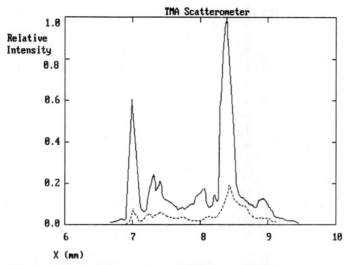

Figure 8.9 Plots of total scatter (solid line) and bulk scatter (dashed line) for a zinc selenide window with bulk defects.

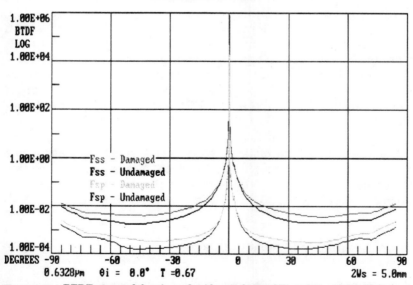

Figure 8.10 BTDF scans of the zinc selenide windows of Figs. 8.8 and 8.9. Taken at 0.633 μm, each window was scanned in the plane of incidence for F_{ss} and F_{sp}.

8.3   Contaminants and Abrupt Surface Defects

The cross-polarization technique can also be used to separate surface-roughness scatter from surface-contaminant scatter. In order to accomplish this, CCD raster measurements (Sec. 6.9) are combined with the separation technique. The result is shown in Fig. 8.11 where colors (shown as shades of gray here) are used to map BSDF scatter values as a function of location on a front-surface mirror. In the top half of the figure, scatter is due predominantly to surface roughness. Notice the prominent scratch running diagonally across the sample. Several dust particles, scattering at about 25 times the background level can also be found. In the lower half of the figure the same sample area is mapped using the separation technique to suppress scatter from surface roughness and pass contaminant-induced scatter. Notice that the same prominent contaminants can be found in the lower image. The contrast between contamination scatter and background has been increased from 25:1 to 1000:1 and is limited by the dynamic range imposed by the CCD camera electronics. Given sufficient dynamic range in the electronics, the contrast ratio should approach the extinction ratio of the polarizers. The sample area in Fig. 8.11 is approximately 6 × 12 mm and the pixel resolution on the sample is 20 μm.

Figure 8.12 shows a BRDF raster map of a front-surface mirror with particulate contamination and polarization filtering. The data are then organized into the histogram shown in Fig. 8.13. BRDF levels are plotted horizontally and the number of pixels at each level vertically. The near-Gaussian shape is due to the noise-distribution signals of near-zero light pixels and the high BRDF spikes are due to contamination.

Figure 8.14 shows the cross-polarization technique used to locate and map splatter defects on a coated-metal surface. A coating has been applied to harden the polished surface. Notice that the scatter signal is weaker on the relatively flat top of the larger defects than on the steeper edges.

8.4   Nontopographic Defects in Opaque Materials

Precision-machined mirrors tend to have much larger cross-polarized BRDFs than their polished counterparts, so one would guess that the amount of postpolishing required to remove subsurface damage from diamond-turned mirrors could be monitored with the cross-polarization technique. The technique can be used with semiconductor wa-

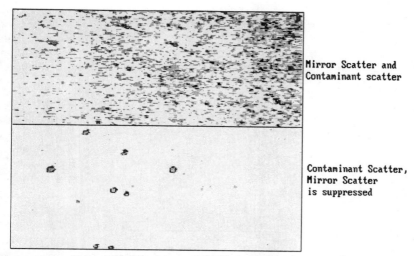

Mirror Scatter and
Contaminant scatter

Contaminant Scatter,
Mirror Scatter
is suppressed

Figure 8.11   The top half is a raster map of all sample scatter. The bottom exposure of the same sample area is a raster map with surface scatter suppressed by the cross-polarization technique.

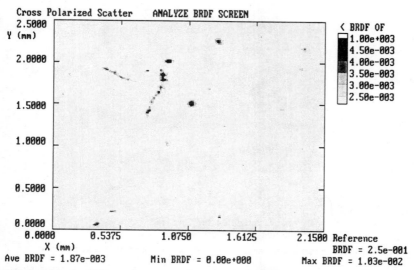

Figure 8.12   A CCD raster scan of a contaminated mirror using the cross-polarization technique.

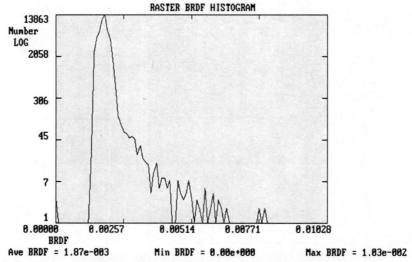

Figure 8.13   A histogram of the raster data in Fig. 8.12. The low BRDF peak is due to noise signals associated with a dark field.

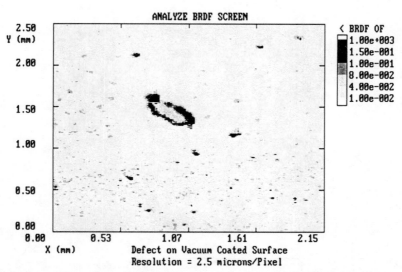

Figure 8.14   The cross-polarization technique used to locate and map coating defects.

fers to identify defects that are within a skin depth or two of the surface. Defects can be located that are not apparent by inspecting the surface of the wafer. This can be especially important in developing high-volume production techniques for materials that are softer than silicon, such as gallium arsenide.

8.5   Summary

The location and mapping of defects using scatter measurements is a powerful inspection technique. Its purpose is often different from more conventional scatter measurements in that the defect size, number, or density may be the issue of concern instead of how much light is scattered. Scatter is just the means of detection and mapping. The cross-polarization technique is a powerful tool in such cases. The true scatter level is suppressed to obtain a low-noise indication of defect location and size. The technique has been applied to a variety of inspection problems and is finding use both inside and outside the optical industry.

Scatter Specifications

The 1970s and 1980s generated considerable concern over scatter in optical systems. Although it was often recognized in advance that low-scatter optics were required for a given application, the specifications were usually either nonexistent (a best effort requirement) or inappropriate. The easiest, most available and cheapest scatter measurement was the TIS measurement. Until the mid-1980s most of the specifications written to handle scatter concerns were either TIS (given without either angle or frequency limits) or rms roughness found from some sort of profile data. Surface roughness was often specified to control scatter, even though it was recognized that it would be difficult, futile, and sometimes impossible to attempt to relate the roughness parameter(s) to actual component scatter. But at least the direction was right (no sign error), as smoother surfaces do generally mean less scatter. Then in the late 1980s, serious work was begun on BSDF standards through an ASTM committee funded in part by the U.S. Air Force. The result is a document that not only details minimum measurement requirements, but also gives a data-format system. This will enable the easy transfer of data between labs and increase the ease with which specifications can be checked—and rechecked. This chapter addresses the problem of selecting the scatter measurement(s) that will make appropriate specifications for a given component, system, or process.

The preceding chapters have presented the definitions and techniques for quantifying, measuring, and analyzing optical scatter. The issue is now approached from the other direction. That is, given that the metrology exists, what specifications should be called out to assure component quality or to provide control of a critical process. This is a key issue. Specifications that are too loose, or too tight, will waste money and time instead of saving them.

Scatter specifications are pretty much the bottom line for this book. You need specifications in order to qualify parts and/or systems. You even need scatter specifications to build a scatterometer. And to be appropriate, specifications need to address the issue at hand—they have to be application specific. Generating the right specification requires knowledge of the system (or process) under design (or test), as well as scatter measurement and interpretation.

This chapter reviews the process of generating appropriate scatter specifications for three classes of problems. Most optics are specified, produced, and sold without knowledge of their eventual application. These require some sort of generic scatter specification which is usually fairly easy to generate. A smaller percentage of optical components is specified to do a particular job. In these cases, the scatter specification should be application specific, and although it may not be easy to generate it, there is usually a well-defined relationship between the specification and system behavior. The idea is to address scatter-sensitive issues in the design phase before they become system problems that require breakthroughs in hardware development. Writing good specifications forces the design effort in the right direction. Several examples are given of conversion from application requirements to scatter specifications. The third class of specifications involves the use of scatter measurements as a means of quality control in situations where there is not a well-defined (or understood) relationship between scatter and the effects causing it. This is often the case in process control applications of scatter measurement. Empirical measurements are often used to define the specification, or in effect, the specification comes well after system design—not before it. The concern is not always with too much scatter, but the knowledge (gained through experience) that a production process is acceptable only if a scatter measurement is above, or below, a fixed value. The following sections review these three cases.

9.1   Generic Specifications

TIS is a good scatter specification to give for general-purpose optics. The corresponding measurement limits and wavelength need to be given with the TIS number. For example, in addition to specifications for reflectance and flatness, front-surface aluminum mirrors could be specified as *having a TIS $< 10^{-3}$ at 0.633 μm over the collection angles between 3 and 85 degrees from the normally incident specular beam.* Without a scatter specification, it is not obvious which components are more suitable for a low-scatter application. This is illustrated in

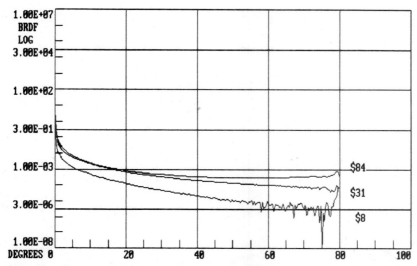

Figure 9.1 Comparison of scatter from three front-surface aluminum mirrors. The price of the mirrors went up with flatness ($\lambda/2$, $\lambda/4$, and $\lambda/10$) and, unfortunately, so did the BRDF.

Fig. 9.1, which shows the BRDF from three front-surface aluminum mirrors. The mirrors were purchased from the same vendor at the prices indicated in the figure. A clear-cut case of getting what you pay for? More BRDF for more money? Actually the mirrors were specified for flatness, not scatter, at ½, ¼, and ¹⁄₁₀ wavelengths. The increase in cost reflected the extra time required to polish the mirror flat—and in the process, increase the scatter. A TIS specification on these mirrors, in addition to the flatness and reflectance values, would have revealed the trend.

This is not an isolated example. It is difficult to purchase off-the-shelf optics that are known to be low scatter. The addition of simple bandpass or antireflection coatings generally increases scatter. Waveplates and polarizers tend to be high-scatter components. One high-scatter element in a transmissive chain of optics can dominate scatter in the system and reduce the expense *required* for other low-scatter components. Manufacturers of generic optics can give scatter specifications as *not to exceed* limits, similar to existing reflectance and flatness specifications, and expressed as piecewise linear plots. This makes a lot of sense for components whose scatter patterns follow apparent power-law distributions (see Secs. 4.5.2 and 7.2.1 on fractals).

Expressing component scatter as a BSDF curve gives considerably more information than TIS measurements. Figure 9.2 shows the BRDF of two flat mirrors at 0.633 µm. The curves have been inte-

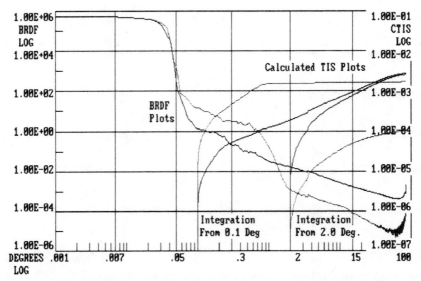

Figure 9.2   The BRDF of two replicated mirrors is compared. *Orange peel* effects on one mirror cause it to have excessive near-specular scatter that would not be caught by TIS measurements, which typically start at about 2 degrees from specular.

grated (and multiplied by a 2πθ factor) to give estimates of the corresponding TIS values over two different sets of limits. Starting the integral near separation from the instrument signature gives an ambiguous result. The TIS values are very close. A real TIS measurement would probably start integration about 2.0 degrees from specular and show a striking difference in quality; however, if the application required low scatter near specular the TIS numbers would result in choosing the wrong component. Two points are clear. Require that angular limits be given with TIS specifications and do not rely on TIS for near-specular scatter requirements. The next section addresses the more difficult issue of reducing system requirements to component specifications.

9.2   Application of Specific Issues and Examples

One of the reasons scatter specifications have not been used extensively (or appropriately) is because it is not always obvious how to get from the functional system requirement to a particular component scatter requirement. Another reason has been the lack of representative data in a timely fashion. Appendix C contains BSDF data for a variety of materials and wavelengths. It is intended to be used as a

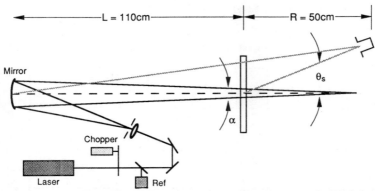

Figure 9.3 Diagram of a scatterometer employing a mirror as the final focusing element. The spatial filter removes most of the source scatter. The scatter specification for the final mirror can be determined from the required instrument signature.

data source that will allow *order of magnitude* and achievable BSDF levels to be used in system design. The following examples illustrate the conversion from real system scatter problems to the corresponding component specifications.

9.2.1   Example 1: scatterometer focusing mirrors

This problem was actually encountered, and solved, in real life by an optical instrumentation company. It may be useful to review the material on instrument signature found in Sec. 6.4. Figure 9.3 is a diagram of a scatterometer that uses a front-surface mirror to focus the incident source beam. The spatial filter removes most of the scatter from the chopper, beamsplitter, and turning mirrors so that the scatter from the source box is dominated by surface defects on the focusing mirror. The receiver optics are arranged so that the focusing mirror leaves the field of view at $\theta_s = 3°$. An early prototype of the instrument used an off-the-shelf 50 cm focal length mirror, that by a stroke of luck, proved to be *low scatter*. The instrument signature for a system to be delivered was quoted on this basis and specified by the piecewise linear representation shown in Fig. 9.4. The instrument under development for the customer required the use of a 30 cm focal length mirror, and unfortunately, these off-the-shelf components, which could not be purchased at the same optical house, proved to be comparatively high scatter. The resulting signatures exceeded the specification by more than an order of magnitude. What is the relationship between instrument signature and mirror scatter, and what should the scatter specification be for the focusing mirror?

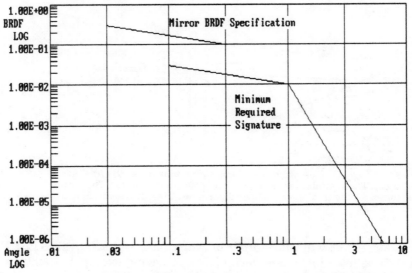

Figure 9.4   Conversion of required instrument signature into a BRDF specification.

Solution. The signature, expressed in BRDF units, is calculated as though it originates from the sample position. By the reasoning presented in Sec. 6.4, the mirror will be completely in the receiver field of view (FOV) for less than a degree. At θ_N (about three degrees in this case) the mirror has left the FOV completely. Thus mirror scatter is reduced to some degree by field of view. An expression can be derived to account for this effect; however, if the mirror meets the specification from 0.1 to 1 degree, it should easily meet it from 1.0 to 3.0 degrees as well. So the question is, what should the mirror BRDF be to meet the 0.1 to 1.0 degree signature specification? The mirror is farther away from the receiver aperture than the sample by the ratio $(1 + L/R)$. Thus the receiver presents a smaller solid angle to the mirror [by $(1 + L/R)^2$] than to the sample. And, by the same reasoning, mirror scatter which deviates by angle α from specular will appear in the signature at location $\theta_s = \alpha(1 + L/R)$. Thus,

$$F_{\text{mir}}[\alpha] = F_{\text{mir}}[\theta_s(1 + L/R)] = (1 + L/R)^2 \, F_{\text{sig}}[\theta_s] \qquad (9.1)$$

Using the dimensions given in Fig. 9.3, the signature requirement, for 0.1 to 1.0 degree, translates up and to the left, as shown in Fig. 9.4, to become the mirror BRDF specification. The straight-line segment can be extended to $\theta_N = 3.0°$ as the required drop in signature will be achieved through controlling receiver FOV. The specification is most

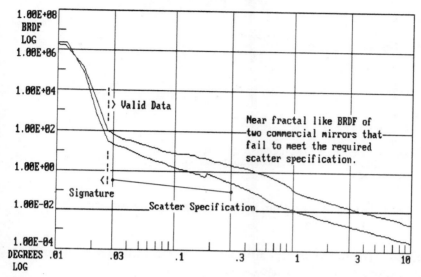

Figure 9.5 Measured BRDF from two mirrors that failed the scatter specification.

easily expressed as shown, in graphical form, or in equation form as

$$\log \left[F_{mir} (\theta) \right] < 0.01 - (1/2) \log \theta \qquad (9.2)$$

$$F_{mir} \left[\theta \right] < 1/\sqrt{\theta} \qquad (9.3)$$

One of the problems with this particular specification was that the optical houses could not check their product. As shown in Fig. 9.5, several mirrors were obtained and measured by the instrumentation company before an acceptable supplier was found.

9.2.2 Example 2: imaging optics

Consider the very simple situation in Fig. 9.6 where a camera (lens focal length f of 6 cm and diameter D of 3 cm) is to be used by an astronaut to image a star located at a small angle θ from the moon. The star is about the brightness of the sun and is 50 light years away. The lens images light from the star and the moon onto separate locations at the image plane. Moonlight, scattered by the lens, creates a glow of light over the entire image plane. Because the moon is so much brighter than the star, a close angle point will exist where the image of the star is lost in the scattered moonlight. The question is how low does the BTDF of the lens have to be in order to photograph the star as

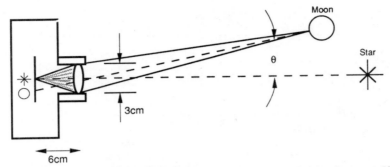

Figure 9.6 The camera is to be used to photograph a star close to the moon. If the moon is too close to the star, moonlight scattered by the lens will obscure the image of the star.

close as 1 degree away? And, could we reasonably expect to image the star closer than 1 degree?

Solution. To illustrate the situation, assume that the minimum acceptable signal-to-noise ratio (starlight density-to-scattered moonlight density) at the image plane is one and the BRDF of the moon is Lambertian in form with a reflectance of 0.1 (i.e., $F_M = 0.1/\pi$). Scattered moonlight intensity at the image of the star is then set equal to the intensity of the imaged starlight in Eq. (9.4). The first two terms give sunlight on the moon in watts. The next two terms convert this to moonlight on the lens and the next two give scattered moonlight on the film, and finally division by A the diffraction-limited area of the star image, gives noise intensity. The right-hand side gives the signal intensity.

$$\left(\frac{P_{\text{sun}}}{\Omega}\right) \Omega_{ms} F_m \Omega_{cm} F_L \Omega_{LA/A} = \left(\frac{P_{st}}{\Omega}\right) \Omega_{S+C/A} \tag{9.4}$$

After cancellation of terms, the required value of the lens BTDF is found to be

$$F_L = \frac{\Omega_{S+C}}{\Omega_{MS}\Omega_{CM}\Omega_{LA}F_M} \approx 30 \ sr^{-1} \tag{9.5}$$

Details of the calculation are given in Fig. 9.7. Figure 9.8 gives the BTDF of a camera lens. The level $30 \ sr^{-1}$ is reached at 0.2 degrees. If the moon moves closer than this to the star, it will be lost from view. If a signal-to-noise ratio of 10 is required, then the maximum BTDF would be $3.0 \ sr^{-1}$ and the star could be photographed no closer than 1.3 degrees from the moon.

This example is very simplistic. It ignores secondary scatter from the image of the moon at the focal plane and from the walls of the

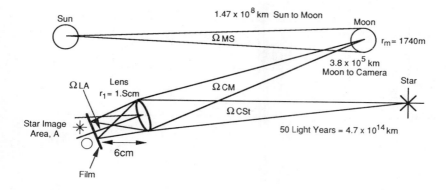

$$\text{Use } F \simeq \frac{P_s}{P_i\,\Omega} \quad \text{and} \quad \Omega = \frac{\text{Area}}{(\text{Distance})^2}$$

$$F_M \simeq \frac{.1}{\pi} = .032 \text{ sr}^{-1} \quad A = \pi \left(\frac{2\lambda f}{\pi D}\right)^2 \quad \text{Equation 6.3 and } \lambda = .6\mu m$$

$$\Omega_{MS} = \frac{\pi\,(1740)^2}{(1.47 \times 10^8)^2} = 4.4 \times 10^{-10} \text{ sr}$$

$$\Omega_{CM} = \frac{\pi\,(1.5 \times 10^{-5})}{(3.8 \times 10^5)^2} = 4.8 \times 10^{-21} \text{ sr}$$

$$\Omega_{LA} = \frac{\pi\,A}{6^2} = 1.6 \times 10^{-9} \text{ sr}$$

$$\Omega_{CSt} = \frac{\pi\,(1.5 \times 10^{-5})}{(4.7 \times 10^{14})^2} = 3.2 \times 10^{-39} \text{ sr}$$

$$\left(\frac{P_{Su}}{\Omega}\right) \Omega_{MS}\, F_M\, \Omega_{CM}\, F_L\, \Omega_{LA}\big/A = \left(\frac{P_{St}}{\Omega}\right) \Omega_{CSt}\big/A$$

$$F_L = \frac{\Omega_{CSt}}{\Omega_{MS}\, \Omega_{CM}\, \Omega_L\, F_M} = 29.6 \text{ sr}^{-1}$$

Figure 9.7 Calculation of lens BTDF.

camera. Practical problems are considerably more complex. But, with very little effort, a scatter specification has been generated that addresses the specific problem at hand. Real space imaging systems often use several reflective elements with complex baffles to reduce scattered light. Analysis requires the use of ray-tracing/scatter-prediction programs to predict critical scatter levels. And the choice of acceptable signal-to-noise ratios depends on the hardware being used (array detectors instead of film, etc.). But the approach is essentially the one outlined here.

9.2.3 Example 3: laser resonator losses

Scatter from laser cavity elements is an unwanted source of intracavity loss. In high-gain lasers, scatter losses are not a signifi-

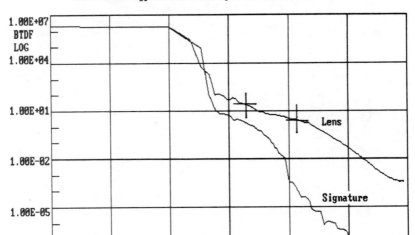

Figure 9.8 The BTDF of the camera lens marked to indicate the 30 $sr^{-1}/0.2°$ and 3 $sr^{-1}/1.3°$ locations that correspond to signal-to-noise levels of 1 and 10, respectively.

cant factor. However, in low-gain, low-power continuous-wave lasers, such as the HeNe lasers used in supermarket checkouts, scatter loss may play a significant role. Relatively inexpensive mirrors are hard sealed onto a relatively expensive tube in the production process. The mirror scatter could be checked before sealing to the tube. The issue here is whether or not scatter from laser-cavity mirrors is of practical concern, and if so, what scatter specification is appropriate?

Solution. A number of authors (Verdeyen, 1989; Siegman, 1986; Yariv, 1976) have presented the development of the simple equation giving the output power in terms of saturation power P_s, the output mirror transmission T, the percent round-trip loss L, and the percent round-trip gain g for low-gain cavities.

$$P_0 = P_s\left(\frac{g}{L + T} - 1\right)\frac{T}{2} \tag{9.6}$$

The gain is proportional to the length of the active medium. Losses are due to scatter and absorption from the beam at the cavity windows and mirrors, and to Rayleigh scatter from gas molecules within the medium. Window losses can be eliminated in some cases by sealing the cavity mirrors directly to the gas discharge tube. For this situa-

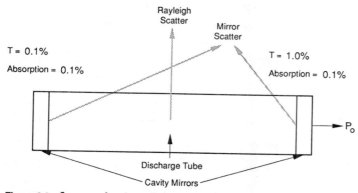

Figure 9.9 Loss mechanisms in a low-power gas laser.

tion, see Fig. 9.9. If the transmission of the output mirror is high, losses can often be ignored entirely because they are small compared to the round-trip reduction in cavity power lost to the output beam. The common route to analyzing Eq. (9.6) is to differentiate with respect to T and demonstrate that there is a value of T which will give maximum output power.

$$T_{opt} = -L + \sqrt{gL} \qquad (9.7)$$

For the shorter, low-power lasers, this value is often around 1 percent, which is comparable with other cavity-loss mechanisms. A slightly different approach is taken here to allow the relative importance of losses to be examined. If the internal round-trip loss is held to about the 1 percent level the round-trip gain is about 4 to 5 percent. Assuming a laser with $T = 1$ percent and $g = 4$ percent the ratio P_0/P_s can be plotted as a function of the round-trip loss as shown in Fig. 9.10. Figure 9.11 shows the measured BRDF for three laser mirrors. One mirror was known to be damaged. The other two were selected from a group of several mirrors to show the spread in BRDF from laser mirrors. The curves have been integrated from 0.75 degrees to grazing to obtain the calculated TIS. Because the mirrors have reflectances of almost 1.0 there is little difference between fractional loss and TIS. The damaged mirror (TIS = 4.8 percent) would probably shut the laser down completely. Two high-scatter mirrors (TIS = 0.23 percent) would reduce laser power by about 30 percent (see Fig. 9.10). The conclusion is that excessive scatter should be of concern. But what specification should be used?

The data cannot be integrated closer to specular because of the instrument signature. However, an estimate can be obtained by extending the curves into the very-near-specular region. This can be done

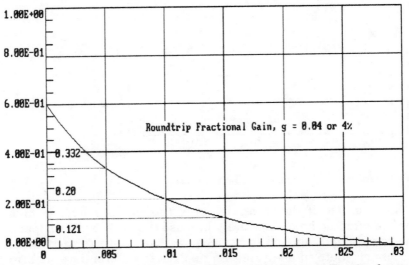

Figure 9.10   Laser output falls as internal cavity losses increase. Fractional scatter losses as high as 0.005 (or 0.5 percent) make a significant difference in laser output power.

graphically (with some software help) or algebraically as follows. Assume that the BRDF is linear on a log-log scale as shown. Then

$$\log F = \log B + M \log \theta \tag{9.8}$$

$$F = B\theta^M \tag{9.9}$$

The fractional loss is obtained by integrating F around the hemisphere.

$$\text{Frac. loss} = 2\pi B \int_{\theta_{min}}^{\pi/2} \theta^{1+M}\, d\theta = \frac{2\pi B}{2+M}\,(\pi/2)^{2+M} - (\theta_{min})^{2+M} \tag{9.10}$$

Values for M and B were obtained from the data of Fig. 9.11. The curves were assumed linear and M was found near specular from the difference between the BRDF values over 1 to 10 degrees. Using this value of slope and the BRDF at 1 degree, B was calculated as the linear-fit BRDF value at 1 radian. These constants were substituted into Eq. (9.10) and the fractional loss calculated. Integration was started at 0.75 degrees to compare to the TIS values. The calculated and measured values agree to within a factor of 2 as shown in Table 9.1, so the conclusion is that our BRDF fit is fairly good. It seems reasonable to

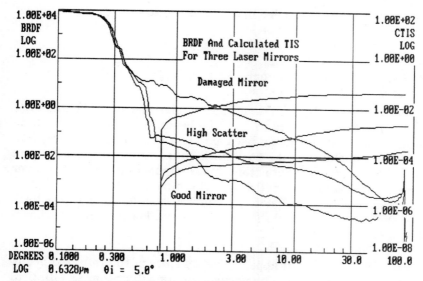

Figure 9.11   BRDF and calculated TIS for three laser mirrors.

TABLE 9.1   Summary of Calculated Scatter Losses

Mirror	B	M	TIS $0.75 \to 90°$	Frac loss $0.75 \to 90°$	Frac loss $0.06 \to 90°$
Damaged	2.13×10^{-3}	-1.8	4.8×10^{-2}	3.6×10^{-2}	4.8×10^{-2}
High scatter	4.4×10^{-4}	-1.2	2.3×10^{-3}	3.8×10^{-3}	3.9×10^{-3}
Good	1.6×10^{-6}	-2.4	2×10^{-4}	1.2×10^{-4}	3.8×10^{-4}

extend the integration into about twice the angular half-width of the diverging output beam (about 1 mrad or 0.06 degree for a typical low-power HeNe laser). This gives the second set of calculated fractional loss values shown in Table 9.1. As expected, if the slope is high a big increase in fractional loss is found, and if the slope is low the difference is much smaller.

Based on the data available here, one would be tempted to call out a specification based on a calculated fractional loss—say less than 0.1 percent over 0.06 to 90 degrees. This will work, but it is rather calculation intensive. If a number of undamaged mirrors that have been coated by the same process are examined, there is a tendency for them to range from the *good* to *high-scatter* mirrors shown in Fig. 9.11. That is, as the BRDF increases the slope tends to decrease. This means that by watching the BRDF at one angle the acceptable mirrors can be quickly found. This trend must be checked for each coating process and an empirical limit decided upon. Thus the specification for this set might be: "The BRDF shall be less than $5 \times 10^{-4} \, sr^{-1}$ at 10 degrees."

9.2.4 Example 4: diffraction from precision-machined turning mirrors

High-power laser systems sometimes make use of large-diameter precision-machined mirrors to turn the beam. It is often desirable to minimize the light diffracted back into the incident beam direction. Even after polishing, some tool-mark diffraction may remain and there have been cases where tool marks reappear on the surface over several months following polishing. The mirror is to be specified in such a way that light will not retrodiffract from the tool marks back into the output laser port of the incident beam. The geometry is shown in Fig. 9.12.

Solution. Diffraction will appear on both sides of the reflected beam. If the mirror is center cut, the various orders will appear as elliptical cones of light with negative orders ($n < 0$) accounting for diffraction back toward the laser port. The two-dimensional grating equations describe the position of the diffracted light in terms of laser wavelength and spatial frequency.

$$\cos \phi_s \sin \theta_s = \sin \theta_i + n f_x \lambda \tag{9.11}$$

$$\sin \phi_s \sin \theta_s = n f_y \lambda \tag{9.12}$$

$$f_x^2 + f_y^2 = f^2 = d^{-2} \tag{9.13}$$

The quantity d is defined as the tool feed, or the distance the tool moves between spindle revolutions. The problem reduces to in-plane considerations only because the closest approach of a diffraction ring to the laser port occurs at $\phi = -180$. Thus from Eq. (9.11)

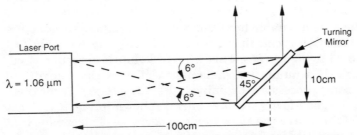

Figure 9.12 Scatter back into the laser port is to be minimized. The geometry dictates that diffraction from the mirror tool marks be eliminated from a 6-degree cone about the incident beam.

$$d = \frac{-n\lambda}{\sin \theta_s + \sin 45°} = 0.75n \Big|_{\theta_s = 45°} \qquad (9.14)$$

where $n = -1, -2, -3,...$ for the problem at hand. Solving for d at $\theta_s = 45°$ gives values of the feed that will diffract light directly back into the laser port. Feeds which diffract within 6° of these directions must also be avoided. These can be evaluated by differentiating with respect to θ_s and setting the differential angle equal to 6 degrees.

$$\Delta d = \frac{-n\lambda\Delta\theta_s}{(\sin \theta_s + \sin 45°)^2} = 0.039n \Big|_{\substack{\theta_s = 45° \\ \Delta\theta_s = 6°}} \qquad (9.15)$$

Feeds must not be used in the ranges given by

$$d + \Delta d = (0.75 \pm 0.039)^n \ \mu m \qquad (9.16)$$

It has been assumed here that the machine tool does not have any prominent internal vibration that produces surface periodicities at other frequencies. If this is not the case, then additional unwanted diffraction will occur as explained in Sec. 4.3 and App. B.

9.2.5   Example 5: scatter in a laser range finder

A junior engineer (the boss's son) has just brought you his design for a laser range finder. This is one job where nothing must go wrong and you need to check it out carefully. The design is shown in Fig. 9.13. The detector samples the outgoing beam by sensing scatter from the beam dump, the beam splitter, and the output window (which has been tilted to avoid direct back reflections). It also senses the return pulse from the target. A microprocessor monitors the time between pulses and calculates the distance to the target. The effective aperture of the detector system is 1 cm² and the distances from the detector to various system elements, with their BRDFs, are found in the chart in Fig. 9.13. The linear response of the detector and a saturation level are given. Once in saturation, the detector is blind for several microseconds and is likely to miss the return pulse. Will the design work?

Solution.   The detector signal from each scatter source may be evaluated in terms of its BRDF as follows.

$$P_s = FP_i\Omega_s \qquad (9.17)$$

Results are given in the following table (Table 9.2). The beam splitter is assumed to be 50 percent reflective.

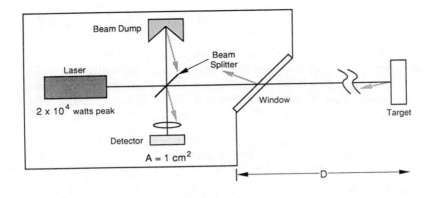

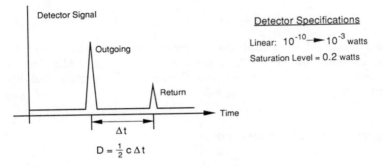

$$D = \frac{1}{2} c \Delta t$$

Object	Distance to Detector	BRDF in Detector Direction
Beam Splitter	5cm	10^{-4} sr^{-1}
Beam Dump	10cm	10^{-3} sr^{-1}
Clean Window	20cm	10^{-4} sr^{-1}
Dirty Window	20cm	10^{-2} sr^{-1}
Target	3100m	10^{-2} sr^{-1}

Figure 9.13 Design features of a simple laser range finder.

TABLE 9.2   Calculation of Scatter Signals at the Detector

Beam splitter	$P_s = 10^{-4}$	$2 \times 10^4 \, (1/5)^2 = 0.08$ watts
Beam dump	$P_s = 10^{-3}$	$0.5 \times 10^4 \, (1/10)^2 = 0.05$ watts
Clean window	$P_s = 10^{-4}$	$0.5 \times 10^4 \, (1/20)^2 = 0.00125$ watts
Dirty window	$P_s = 10^{-2}$	$0.5 \times 10^4 \, (1/20)^2 = 0.125$ watts
Target	$P_s = 10^{-2}$	$0.5 \times 10^4 \, (1/3.1 \times 10^5) = 5 \times 10^{-10}$ watts

The system will not work when the window gets dirty. However, if the peak laser power is cut back to 10^4 watts, the system will function at distances beyond 3100 m. The question of whether or not to completely redesign your future boss's system, with the illuminated window section out of the detector field of view (which avoids the issue), is well beyond the scope of this book.

9.3   Process Control Issues and Examples

The object of scatter measurement for process control is often not the BRDF of the sample, but the indication of change on or in a material. Samples are often diffuse and may be in hard-to-reach hostile environments. It is the noncontact, real-time advantages of scatter measurements that are employed. For example, particulate emissions from an exhaust system can be monitored. Because scatter is not the key issue, it is more difficult to obtain a direct scatter specification. Often an empirical relationship between process quality and the measured scatter needs to be developed. In the case of the exhaust stack, it might be learned by experience that if the BTDF at 20 degrees rises above 10^{-3} sr^{-1} the downwind neighbors will start complaining to county officials about the stack odor. The following examples illustrate this growing use of scatter metrology.

9.3.1   Detection of paper flaws

The paper industry currently uses scatter as a means of process control. Continuous sheets, or webs, of paper which are 1 to 5 m wide, require inspection for holes, blotches, streaks, and coating nonuniformities. At web speeds up to 2000 m/min, a streak can scrap a lot of paper in a short time. Two inspection techniques are commonly used (Paumi, 1988). The laser technique, which is shown in Fig. 9.14, consists of a laser scanner used to produce a line of light across the moving web. A detector is then placed in the resulting transmissive (or reflective) scatter pattern and the signal is monitored to check for defects under computer control. If 100-percent coverage of the paper is required, the web speed is limited by scan rate, spot size, and detector sensitivity. There is obviously a design tradeoff between minimum defect size (often less than 1 mm) and speed. Depending on the sophistication of the device, the system may be capable of discriminating between the various kinds of defects. A streak may stop the paper, while holes or blotches may have to reach a critical density or size before

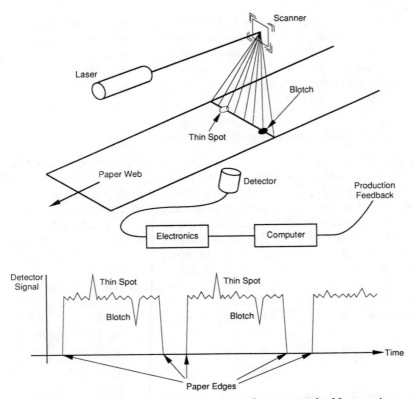

Figure 9.14   Transmissive scatter used to detect flaws on a web of fast moving paper.

production is stopped. Figure 9.15 compares the BRDF from paper with three different gloss coatings. Notice that the three have distinctly different BRDF and that all of them are very flat at high-scatter angles. It is this flat characteristic, which is typical of many rough (diffuse) surfaces, that is exploited in this process control system. By placing the detector well away from the specular direction, it is insensitive to changes in scatter direction and incident direction. A second type of system employs an array camera and a white light source. Data are taken in a manner similar to the fast raster scans described in Sec. 6.9. Although the CCD camera instruments are less expensive, they are slower for a given resolution.

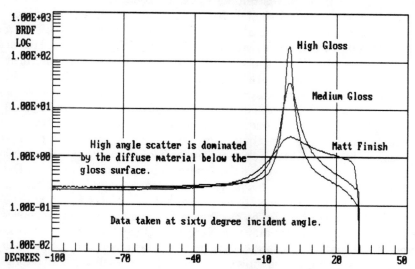

Figure 9.15 BRDF of three different papers with different clay coatings to produce gloss. High-angle scatter is dominated by the diffuse surface under the coating.

9.3.2 Noncontact monitoring of emissivity and temperature

A number of industrial processes involve first heating materials to several hundred degrees centigrade and then cooling them in a controlled manner to bring about desired material changes. Temperature control of the process to a few degrees can be critical. Because the processes often involve molten (or near-molten) materials, monitoring the temperature can be difficult. Techniques which have been developed to calculate temperature (DeWitt, 1986; Tanka and DeWitt, 1989) based on measurements of infrared radiation from the material require that the emissivity be known. Unfortunately the emissivity varies not only with material type, wavelength, and temperature, but also with surface roughness—and the roughness can change dramatically during some of these processes.

An example is the galvanneal process used to improve the quality of galvanized steel. In a hot-dip galvanization process a coating of zinc several micrometers thick is applied to cold-rolled steel to prevent corrosion and improve paint adhesion. The coating properties can be improved if the coated steel is heated further, allowing the zinc to diffuse into the steel. This requires temperatures of about 600°C and control to about ±10°C. The surface changes visibly from a rather shiny molten zinc to a duller, rougher alloy finish as the material is cooled. In work initiated at Purdue University (DeWitt and Nutter, 1989; Hill et al., 1989), the BRDF of the surface will be measured to monitor changes in surface structure and allow the effects of changes in emissivity to be entered into the temperature calculation. A well-controlled, repeatable process can make a big difference in quality and economics to high-volume users such as the automobile industry.

Because the surface can change drastically in finish during these processes, it is often difficult to meet the *optically smooth* requirement for calculation of surface statistics. And depending on the process, the *front-surface* criteria may also be violated at some point. The inherent limits on the ability to directly calculate surface statistics should be established for each process by taking BRDF measurements on representative samples at various wavelengths. Once the limits are exceeded, it will not be possible to rely on the relationships of Chap. 4 to find the surface roughness; however, the BRDF is still a sensitive indication of changes in surface roughness. By measuring the BRDF, empirical relationships can be developed, specific to a given process, that will allow surface finish (and hence emissivity) changes to be monitored.

Figures 9.16 and 9.17 show the BRDF at two different wavelengths of several steel plates that have completed different phases of the galvanneal process. At 10.6 μm a specular peak is still plainly observed even though the surface is rough to the eye. This is not the case at 1.06 μm as shown in Fig. 9.17. The longer wavelength is a better indicator over the full length of the process.

A second example is the manufacture of ball bearings. There is an optimum surface roughness. Too rough a surface increases wear and too smooth a surface limits the ability to hold lubrication. There is interest in manufacturing ball bearings in space where globs of molten metal will form good spherical shapes. The surface roughness is determined by the rate at which they cool. In order to properly develop the heating and cooling processes, many of the problems encountered in the galvanneal example above are encountered.

The BRDF measurement of small-diameter spheres presents its own problems. The geometry is shown in Fig. 9.18. In principal it would seem that if the incident beam converges toward a focus located at the center of the sphere, the beam will reflect directly back on itself with only scattered light outside the incident beam volume. Unfortunately, physical optics is a little more subtle than geometrical optics. If a TEM_{00} laser source is used, which allows a tight beam, then the light is easily described by a Gaussian beam (Verdeyen, 1989). The beam

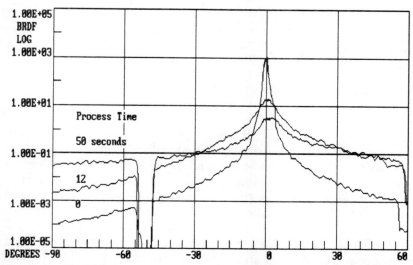

Figure 9.16   The BRDF at 10.6 μm becomes progressively more diffuse a function of process time.

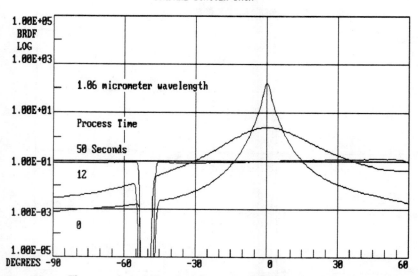

Figure 9.17   The samples of Fig. 9.16 are remeasured at roughly one-tenth the wavelength. At 50 s the surface looks nearly Lambertian at this wavelength.

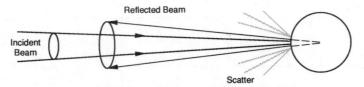

Figure 9.18 The BRDF measurement of reflective surfaces.

never reaches a point focus. It has a minimum spot size where the phase surface is flat (infinite radius of curvature). At beam positions near focus the phase surface can be approximated by a large radius sphere. The result is that the specularly reflected light is a cone of considerably larger diameter than the incident beam. The smaller the diameter of the sample the more pronounced is the effect. Figure 9.19 shows the BRDF of two ball bearings taken at wavelengths of 0.633 μm and 1.06 μm. The larger-diameter ball has a smaller reflected specular beam. Scatter well away from specular is reasonably flat and is higher than most optical surfaces.

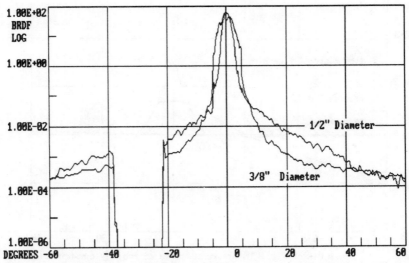

Figure 9.19 BRDF of two ball bearings. The incident angle was 15 degrees and the wavelength, 0.633 μm.

9.4 Summary

Appropriate scatter specifications are the key to obtaining economic advantages from scatter metrology. For many *generic* uses a TIS or simple *not to exceed* BSDF limit makes perfect sense. Vendor specifications are set at the levels they are able to economically maintain. For optics that are purchased to be used in critical low-scatter applications and for the designers of those systems, application-specific specifications are needed. The specification must be tight enough to guarantee system performance, but not so overly tight that the design or component cost is unnecessarily affected. These specifications are harder to generate than the generic ones and require that someone knowledgeable in both scatter and the product design consider these issues at an early development stage. The economic advantages for properly setting these specifications can be large. It costs a lot of money to put a high-scatter mirror into space—only to find out that it has caused the failure of the entire project. Harder yet are the specifications required for process control applications where product quality parameters are related to scatter only by experience or the generation of an empirical relationship. However, for many high-volume industrial applications, the economic benefits can be very significant.

The ability to write specifications that are both technically and economically sound is a sign of expertise in a given field. Problems with writing good specifications should naturally lead to the key questions or to the missing pieces of required information. The people who can write scatter specifications that are appropriate for a company's product become *the company scatter experts*—which brings us full circle to the goals stated in the preface.

Review of Electromagnetic Wave Propagation

Some sections of the book rely on the reader's familiarity with various aspects of electromagnetic field theory. This appendix briefly reviews wave propagation, the idea of a complex refraction index, the Poynting vector, and the diffraction limit. The concepts introduced here are only reviewed—not fully developed. Maxwell's equations are used as the starting point. SI units (meter, kilogram, second, ampere) are used throughout.

A.1 The Wave Equation

Assuming that there are no external free charges or currents, Maxwell's equations can be written in terms of the electric-field intensity **E**, the electric-displacement vector **D**, the magnetic-flux density **B**, and the magnetic-field intensity **H**, as

$$\nabla \times \mathbf{E} = -\frac{\delta \mathbf{B}}{\delta t} = -\mu \frac{\delta \mathbf{H}}{\delta t} \tag{A.1}$$

$$\nabla \times \mu \mathbf{H} = \nabla \times \mathbf{B} = \mu \sigma \mathbf{E} + \mu \in \frac{\delta \mathbf{E}}{\delta} \tag{A.2}$$

$$\nabla \cdot \mathbf{D} = \nabla \cdot \in \mathbf{E} = 0 \tag{A.3}$$

$$\nabla \cdot \mathbf{B} = 0 \tag{A.4}$$

D, **E**, **B**, and **H** are in bold to indicate that they are vector quantities. The symbols μ, σ, and $\in$ represent the medium permeability, conductivity, and dielectric constants, respectively. Taking the curl of Eq. (A.1) and substituting Eq. (A.2) to eliminate **B** (or **H**) gives

$$\nabla \times (\nabla \times \mathbf{E}) = -\frac{d(\nabla \times \mathbf{B})}{dt} \tag{A.5}$$

Using the identity

$$\nabla \times (\nabla \times \mathbf{E}) = \nabla(\nabla \cdot \mathbf{E}) - \nabla^2\mathbf{E} = -\nabla^2\mathbf{E} \tag{A.6}$$

gives the differential relationship

$$\nabla^2\mathbf{E} = \mu\sigma\frac{d\mathbf{E}}{dt} + \mu \in \frac{d^2\mathbf{E}}{dt^2} \tag{A.7}$$

which is known as the wave equation. An identical equation may be found for $\mathbf{B}$ by eliminating $\mathbf{E}$. Solutions for $\mathbf{E}$ in different mediums will now be found.

A.2 Electromagnetic Plane Waves in Free Space

In free space

$$\mu = \mu_0$$

$$\mathbf{E} = \mathbf{E}_0 \tag{A.8}$$

$$\sigma = 0$$

One possible solution to Eq. (A.7) can be shown to take the form of

$$\mathbf{E} = \mathbf{E}_0 e^{j(2\pi\nu\sqrt{\mu_0 \in_0}\, z - 2\pi\nu t)} \tag{A.9}$$

where $\mathbf{E}_0$ is a constant vector that determines electric-field amplitude and polarization direction. The parameter ν is the frequency of the sinusoidal wave and $1/\sqrt{\mu_0\in_0}$ is identically the speed of light c in a vacuum. The usual convention of writing the solution in terms of a complex phaser, but recognizing that only the real part is of interest, has been used. The specific solution shown in Eq. (A.9) is a wave propagating in the z direction. The more general solution is given in terms of the propagation constant k, which is the phase increase per unit propagation distance, and is defined as

$$k = 2\pi\nu/c = 2\pi/\lambda \tag{A.10}$$

The propagation constant is also defined as a vector $\mathbf{k}$ of magnitude k in the direction perpendicular to surfaces of constant phase. Then

$$\mathbf{E} = \mathbf{E}_0 e^{j(\mathbf{k} \cdot \mathbf{r} - 2\pi\nu t)} \tag{A.11}$$

The full solution to Eq. (A.7) is actually the summation of many waves of the form of Eq. (A.11) plus their complex conjugates. If $k < 0$ then the wave travels in the opposite direction. Some texts define plane waves with the negative of the exponent shown in Eq. (A.11). This apparent difference is resolved when the real part is taken. An identical solution set exists for **B**. The two field vectors can be shown to be perpendicular to each other and to **k**, making the solution a transverse wave. Figure A.1 shows the relative directions of **E**, **B**, and **k** (or **S**) for the solution.

Substituting the plane-wave solution into Maxwell's equations and manipulating gives a relationship for η_0, the impedance of free space which evaluates to 377 ohms. This expression can be used for other mediums by substituting the appropriate material constants.

$$\eta_0 = \frac{|\mathbf{E}|}{|\mathbf{H}|} = \frac{2\pi\nu\mu_0}{k} = \frac{k}{2\pi\nu\epsilon_0} = \sqrt{\frac{\mu_0}{\epsilon_0}} \approx 377 \text{ ohms} \qquad (A.12)$$

The Poynting vector **S** gives the instantaneous power density (watts per unit area) associated with the wave. For isotropic media it has the same direction as **k**. In much of the literature, time average power density is expressed as the scaler I and that notation is used throughout this book. For sinusoidal fields the time average introduces a factor of ½. The resulting equations are analogous power calculations based on Ohm's law.

$$\mathbf{S} = \mathbf{E} \times \mathbf{H}^* \qquad (A.13)$$

$$I = \frac{1}{2}|\mathbf{E} \times \mathbf{H}| = \frac{1}{2}\frac{|\mathbf{E}|^2}{\eta_0} = P/A \qquad (A.14)$$

The * (asterisk) indicates that the complex conjugate is taken. P is the power measured over cross-sectional area A.

A true plane wave has an infinite transverse width and no beam divergence (angle spread). This makes sense because with infinite width there is no room for divergence. However, beams of finite width do diverge. The case of a plane-wave incident upon a limiting aperture is

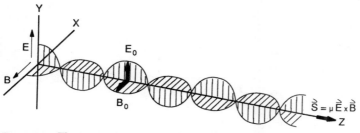

Figure A.1   The transverse nature of the EM wave. The wave is plotted in space for an instant of time.

covered in Chap. 3 on diffraction. The common situation of a finite-width laser beam with a Gaussian electric-field cross section is analyzed in many texts (Verdeyen, 1989; Yariv, 1976) and the results are useful for developing the practical measurement applications described in Chap. 6. Gaussian beams have electric-field cross sections that are described by

$$E = E_0 \frac{\omega_0}{\omega(z)} e^{-[r/\omega(z)]^2} e^{j[kz - \tan^{-1}(z/z_0) + [kr/2R(z) - 2\pi\nu t]}$$ (A.15)

where

$$\omega^2(z) = \omega_0{}^2\left[1 + \left(\frac{z}{z_0}\right)^2\right] \equiv e^{-1} \qquad \text{beam radius}$$

$$R(z) = Z\left[1 + \left(\frac{z_0}{z}\right)^2\right] \equiv \text{phase radius of curvature}$$

$$z_0 = \frac{\pi\omega_0{}^2}{\lambda} \equiv \text{characteristic length}$$

The geometry, shown in Fig. A.2, is for a beam propagating in the z direction. The beam has an e^{-1} field radius of $w(z)$ that has a minimum width ω_0 located at $z = 0$. The beam radius expands to $2\omega_0$ after traveling distance z_0. Cross-sectional amplitude variations are described by the first three terms in Eq. (A.15). The second exponential term contains the phase information. At $z = 0$, $R(z) = \infty$ and it reduces to our phase description of a plane wave. Notice that knowledge of the wavelength and either ω_0 or z_0 is enough to define everything about the beam except total power. For example, it can be shown that the beam radius expands to approach the asymptotic limits defined by θ_{div} as shown in Fig. A.2. For visible wavelengths divergences are small—approximately a milliradian for a conventional HeNe laser. The mini-

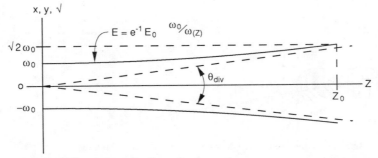

Figure A.2 Divergence of a Gaussian beam.

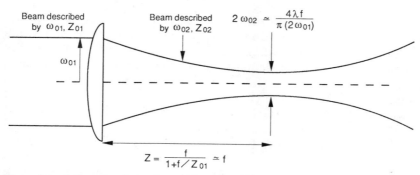

Figure A.3 A Gaussian beam focused by a lens.

mum focused spot size can be calculated as shown in Fig. A.3. A broad (slowly diverging) Gaussian beam $[\omega(z) = \omega_{01}]$ is focused by a thin lens to a diffraction-limited spot diameter of $2\omega_{02}$ located approximately one focal length from the lens.

Beam divergence and minimum spot size are realities that must be dealt with in the design of optical instrumentation. As indicated in Chap. 6, the width of the focused source beam in a scatterometer limits the largest measurable value of BSDF, and divergence limits the ability to work with long thin beams—especially in the infrared. However, the plane-wave approach to analyzing wave behavior is a useful tool and the results are indicative of the behavior expected in many practical situations. The next two sections analyze wave behavior in dielectrics and metals assuming plane-wave propagation.

A.3 Plane Waves in a Dielectric

In a nonmagnetic, nonconducting dielectric, such as glass

$$\mu = \mu_0$$

$$\in \; = \; \in_0 \in_r \tag{A.16}$$

$$\sigma = 0$$

The velocity of light is now given by

$$v = \frac{1}{\sqrt{\mu_0 \in_0 \in_r}} = \frac{c}{\sqrt{\in_r}} = c/n \tag{A.17}$$

where n is defined as the index of refraction. The propagation constant (phase change/unit distance) increases to

$$k = \frac{2\pi v}{v} = \frac{2\pi v \sqrt{\in_r}}{c} = \frac{2\pi n}{\lambda} = \frac{2\pi}{\lambda_m} \tag{A.18}$$

where λ_m is the shortened wavelength in the dielectric medium. The wave still propagates according to Eq. (A.11), but with the new value of k. The physical explanation for the change in velocity lies in the polarization of the dielectric material. The dipoles formed by displacing the bound electrons in the dielectric are set into vibration by the incident wave. The moving charges reradiate at the same frequency, but with a slightly retarded phase when compared to the incident beam. The resultant field is the sum of the reduced-incident field and the polarization-radiated field. Because slightly retarded waves are combining with the beam at every point within the medium, and because sinusoids of a fixed-frequency (but different phase) sum to a sinusoid of the same frequency (but intermediate phase), the resultant field propagates at a lower net velocity.

There is some loss of beam power as it propagates through the dielectric. This occurs for two reasons. First, there are some scattering losses from the bulk material. A small fraction of the reradiated field propagates in directions different from that of the incident field. This power is lost to the beam, but is not strongly absorbed by the dielectric. Second, a small amount of power will be absorbed by the dielectric. This is sometimes thought of as being due to viscous damping of the bound electrons. That is, the dipoles don't quite swing all the way back. Power losses in a material are usually expressed as an exponential decay by means of a loss coefficient α with units of inverse length. For propagation in the z direction

$$I = I_0 e^{-\alpha z} \tag{A.19}$$

Equation (A.19) can be arrived at intuitively by solving the linear differential equation that describes the loss in I per unit propagation distance as proportional to I. However, it can be found directly for the case of absorption losses, by considering that the material has a complex dielectric constant.

$$\hat{\in} = \in_0 \hat{\in}_r = \in_0 (\in_r + j \in_r') \tag{A.20}$$

where for dielectrics $\in_r \gg \in_r'$ and $\Psi = \tan^{-1}(\in_r'/\in_r)$ is a small angle. The implication is that k and n are also both slightly complex. Substituting into Eqs. (A.18), (A.11), and (A.19) gives

$$\hat{k} \simeq \underbrace{\frac{2\pi\nu}{c} \sqrt{\in_r} \cos \Psi}_{\simeq\, n} = j \underbrace{\frac{2\pi\nu}{c} \sqrt{\in_r} \sin \Psi}_{\alpha/2} \tag{A.21}$$

$$E = E_0\, e^{-\alpha z/2}\, e^{-j[(2\pi\nu|n|z/c) - 2\pi\nu t]} \tag{A.22}$$

$$I = \frac{E_0{}^2}{\eta} e^{-\alpha z} \quad \text{and} \quad \eta \simeq \sqrt{\frac{\mu_0}{\epsilon_0\,\epsilon_r}} = \eta_0 n \qquad (A.23)$$

for propagation in the z direction. The first exponential term in Eq. (A.22), which is due to the imaginary term in the dielectric constant, accounts for the losses in the material. For glass and a wavelength of 0.633 μm, α is on the order of 0.005 mm^{-1}. That is, it takes tens of centimeters before beam intensity is reduced by e^{-1}. The second exponential term in Eq. (A.22) describes phase propagation in the medium and is identical to that for the lossless dielectric.

When the dielectric is isotropic (as in glass), and polarization occurs as easily in one direction as in another, the material has only one index of refraction and the above equations provide an accurate description of propagation. However, many materials are anisotropic in nature (quartz, mica) and have one (or two) directions of preferred polarization. Then the index of refraction depends on the direction of polarization and the material is said to be birefringent. In these cases, the dielectric constant can take on several direction-dependent values and is described by a tensor. If propagation is not along, or perpendicular to, a preferred direction, then $\mathbf{E}$ and $\mathbf{k}$ are no longer perpendicular ($\mathbf{S}$ and $\mathbf{k}$ are not parallel). For very high field strengths, such as those found in some pulsed lasers, the induced-material polarization will not be proportional to the incident field. This gives rise to the field of nonlinear optics, which includes effects such as optical harmonic generation and Raman scattering. Although these two effects, birefringence and optical nonlinearity, are important in some scatter measurement situations, they are beyond the scope of this review.

A.4 Plane Waves in a Conducting Medium

In a conductor

$$\mu = \mu$$
$$\epsilon = \epsilon_0\,\epsilon_r \qquad (A.24)$$
$$\sigma > 0$$

Both terms on the right side of Eq. (A.7) must now be considered. When a solution of the form of Eq. (A.11) is substituted into Eq. (A.7), it is found that the propagation constant and the frequency are defined by a dispersion relation

$$\hat{k}^2 = (2\pi\nu)^2 \mu\epsilon + j2\pi\nu\mu\sigma \qquad (A.25)$$

If the frequency is allowed to be complex the fields will damp in time, which does not pertain to the steady-state solutions being reviewed here. Thus to satisfy Eq. (A.25), it is again necessary for k to be complex. Using a little foresight $\hat{k}$ is defined in terms of the two real numbers α and β as

$$\hat{k} = \beta + j\alpha/2 \tag{A.26}$$

Then, as before,

$$E = E_0\, e^{-\alpha z/2}\, e^{j(\beta z - 2\pi\nu t)} \quad \text{and} \quad I = I_0\, e^{-\alpha z} \tag{A.27}$$

Thus, the effect of conduction losses is to introduce a damping term, just as in the lossy dielectric. The skin depth $1/\alpha$ is on the order of a few hundred angstroms (or less) for metals and a few thousand angstroms for semiconductors.

The quantities $\alpha/2$ and β can be evaluated by substituting Eq. (A.26) into Eq. (A.25) and equating real and imaginary parts.

$$\beta = \frac{2\pi}{\lambda}\left[\sqrt{1 + \left(\frac{\sigma}{2\pi\nu\in}\right)^2} + 1\right]^{1/2} \tag{A.28}$$

$$\alpha/2 = \frac{2\pi}{\lambda}\left[\sqrt{1 + \left(\frac{\sigma}{2\pi\nu\in}\right)^2} - 1\right]^{1/2} \tag{A.29}$$

It is common to view conduction losses as caused by a complex dielectric constant or, equivalently, a complex index of refraction. To see how this comes about, consider the definition of the propagation constant. If $\hat{k}$ is complex, as required in the above analysis, then $\hat{v}$, $\hat{\lambda}_m$, $\hat{n}$, and $\hat{\in}_r$ are also complex.

$$\hat{k} = 2\pi/\hat{v} = 2\pi/\hat{\lambda}_m = 2\pi\hat{n}/\lambda = 2\pi\sqrt{\hat{\in}_r}/\lambda \tag{A.30}$$

This means that the complex index can be expressed in terms of $\alpha/2$ and β.

$$\hat{n} = \frac{\lambda}{2\pi}\hat{k} = \frac{\lambda}{2\pi}[\beta + j\alpha/2] \tag{A.31}$$

$$\hat{n} = n + jnK = n + jK_0 \tag{A.32}$$

The complex index is usually defined in terms of the real index and the absorption index K or the absorption coefficient K_0, which are known as the optical constants of metals. Notice that these real constants depend on the conductivity. Values for n and nK are commonly found by a process known as ellipsometry. This involves measuring the reflectance of s- and p-polarized light at the angle of incidence corresponding to minimum p reflectance (similar to Brewster's angle). If the index of refraction is complex, then the dielectric constant is also. Rewriting Eq. (A.25) allows the complex dielectric to also be defined in terms of the conductivity and the optical constants.

$$\hat{k}^2 = (2\pi\nu)^2 \, \mu \left(\in + j \, \frac{\sigma}{2\pi\nu} \right) \qquad \text{(A.33)}$$

$$\underbrace{\phantom{\in + j \frac{\sigma}{2\pi\nu}}}_{\hat{\in}}$$

$$\hat{\in} / \in_0 \, = \, \in_r + j \, \frac{\sigma}{\in_0 2\pi\nu} = \hat{n}^2 = n^2 - (nK)^2 + jzn^2K \qquad \text{(A.34)}$$

Once the optical constants are known at the wavelength of interest (from experiment or handbook), Eq. (A.34) can be used to provide the complex dielectric constant required for computation of the polarization constant Q, described in Chap. 5.

B

Kirchhoff Diffraction from Sinusoidal Gratings

The objective of this appendix is to review a scaler Kirchhoff calculation of diffraction from surfaces composed of sinusoidal reflection gratings by the method outlined in Sec. 3.2. The surface material is assumed to have a reflectance of 1.0. The examples are important because of the use of Fourier composition to represent more arbitrary surface topography. Equation (3.28) will be used. This requires that the Fourier transform of $E_a(x,y,0)$, the incident field in the aperture, be found.

Figure B.1 shows light incident at angle θ_i upon a sinusoidal surface of amplitude a, frequency f_1, and phase α that is propagating in the x

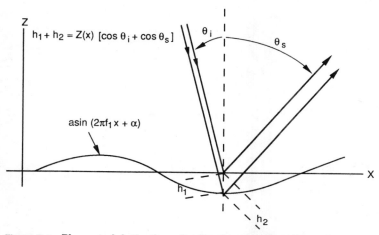

Figure B.1 Phase modulation by reflection from a sinusoidal surface.

direction. Ray 2, diffracted from the actual surface at angle θ_s, is compared to ray 1, diffracting at the same angle from the mean ($z = 0$) surface. These rays will interfere at infinity and thus meet the *far-field* conditions required in the development of Eq. (3.28). Because of the unit reflectance assumption there is no amplitude modulation of the reflected rays. The path length ($h_1 + h_2$) is the extra distance traveled by ray 2 and it imposes a phase difference $\Delta(x)$ between rays 1 and 2 that is proportional to $z(x)$. This amounts to phase modulation (without amplitude modulation) of the wave at the aperture.

$$\Delta(x) = \frac{2\pi(h_1 + h_2)}{\lambda} = k(\cos\theta_i + \cos\theta_s)z(x) \tag{B.1}$$

For the sinusoidal grating

$$\Delta(x) = \underbrace{ka\,(\cos\theta_i + \cos\theta_s)}_{\Delta}\sin\,(2\pi f_1 x + \alpha) = \Delta\sin\,(2\pi f_1 x + \alpha) \tag{B.2}$$

where Δ is the peak value of $\Delta(x)$.

The work of Secs. 3.1 and 3.2 treats diffraction from transmissive apertures. Figure B.2 shows how we can take advantage of this for the case of reflective samples. The plane-wave source, which is actually viewed in reflection, may be thought of as coming from its mirror image. The square grating of side L becomes a square aperture with a

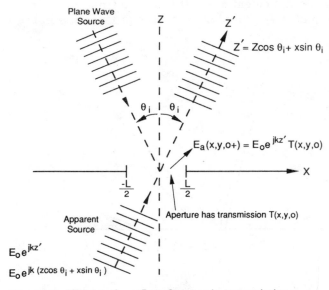

Figure B.2 Viewing the reflected source in transmission.

transmission $T(x,y,0)$ that induces phase modulation.

$$T(x,y,0) = e^{jk\Delta(x)} \text{ rect } (x/L) \text{ rect } (y/L) \qquad \text{(B.3)}$$

where

$$\text{rect } [x/L] = 1 \text{ for } |x| \leq L/2 \qquad \text{and} \qquad = 0 \text{ for } |x| > L/2$$

The plane-wave source propagates along the z' axis in the xz plane at an angle θ_i to the z axis. A rotational coordinate transformation is used to evaluate the source wave at the $z = 0$ plane for any incident angle, θ_i, just as it passes through the aperture (see Fig. B.2). The wave just following the aperture is then given by the rotated incident wave times the aperture transmission, given in Eq. (B.3).

$$E_a(x,y,0) = E_0\, e^{j[kx \sin \theta_i + \Delta \sin (2\pi f_1 x + \alpha)]} \text{ rect } (x/L) \text{ rect } (y/L) \qquad \text{(B.4)}$$

The following identity is used to rewrite the aperture field in a form more convenient for integration.

$$e^{j\Delta \sin \Phi} = \sum_{n = -\infty}^{\infty} J_n(\Delta)\, e^{jn\Phi} \qquad \text{(B.5)}$$

where Δ is constant and Φ is variable. $J_n(\Delta)$ is an nth-order Bessel function of the first kind. Remember that $\sin \theta_i$ in Eq. (B.4) is a constant. Then

$$E_a(x,y,0) = E_0 \sum_{n = -\infty}^{\infty} J_n(\Delta)\, e^{j[kx \sin \theta_i + n2\pi f_1 x + n\alpha]} \text{ rect } (x/L) \text{ rect } (y/L) \qquad \text{(B.6)}$$

Substitution into Eq. (3.28) gives an expression which requires that the Fourier transform of the aperture field be found.

$$E(x_s,y_s) = \frac{\cos \theta_s}{j\lambda R}\, e^{jk[R + (x_s^2 + y_s^2)/2R]} \int\limits_{-\infty}^{\infty}\!\!\int E_a(x,y,0)e^{-j2\pi(f_x x + f_y y)}\, dxdy \qquad \text{(B.7)}$$

The transform may be done by inspection if four points are remembered.

1. $\mathscr{F}[e^{j2\pi(f_1 x + f_2 y)}] = \delta(f_x - f_1, f_y - f_2)$

where $\mathscr{F}$ denotes the Fourier transform and $\delta(0,0)$ an impulse function at $(0,0)$.

2. $[\text{rect } (x/L) \text{ rect } (y/L)] = L^2 \text{ sinc } (Lf_x) \text{ sinc } (Lf_y)$

3. The Bessel function argument Δ is not a function of x and y.

4. Multiplication in the space domain is convolution (given by *) in the frequency domain.

$$E_a = E_0[L^2 \text{ sinc } (Lf_x) \text{ sinc } (Lf_y)]*$$

$$\times \left[\sum_{n=-\infty}^{\infty} J_n(\Delta) e^{jn\alpha} \delta\left(f_x - nf_1 - \frac{\sin\theta}{\lambda}, f_y \right) \right] \quad (B.8)$$

The sinc functions are narrow and the convolution simply acts to impose their shape on the impulse functions. That is, the diffracted orders will have sinc^2 cross sections.

$$E_a = E_0L^2 \sum_{n=-\infty}^{\infty} e^{jn\alpha} J_n(\Delta) \text{ sinc }\left[L\left(f_x - nf_1 - \frac{\sin\theta_i}{\lambda} \right) \right] \text{ sinc } [Lf_y] \quad (B.9)$$

The impulse functions (and the sinc functions) have nonzero values only for $f_y = 0$ and

$$f_x = (\sin\theta_i)/\lambda + nf_1 \quad (B.10)$$

Substituting $f_x = x_s/\lambda z$, $z = R$, $x_s/R \simeq \sin\theta_s$ and rearranging terms gives

$$\sin\theta_s = \sin\theta_i + nf_1\lambda \quad (B.11)$$

which is the grating equation (introduced in Chap. 1). The approximation sign is the result the small-angle approximations used to derive Eq. (3.28). Substituting Eq. (B.8) into Eq. (B.6) and expressing the spatial frequencies on the xy plane in terms of position on the observation plane gives

$$E(x_s, y_s) = \frac{E_0L^2 \cos\theta_s}{j\lambda R} e^{jk\left[R + \left(\frac{x_s^2 + y_s^2}{2R} \right) \right]} \sum_{n=-\infty}^{\infty} e^{jn\alpha} J_n(\Delta) \cdot$$

$$\text{sinc }\left[L\left(\frac{x_s}{\lambda R} - nf_1 - \frac{\sin\theta_i}{\lambda} \right) \right] \text{ sinc }\left[\frac{Ly_s}{\lambda R} \right] \quad (B.12)$$

Equation (B.12) is easily squared if the sinc functions are narrow compared to their spacing. This is normally true and merely implies that many spatial wavelengths are present within the grating, or $f_1 \gg 1/L$. Then

$$I(x_s, y_s) = \frac{1}{2\eta} \left(\frac{E_0L^2 \cos\theta_s}{\lambda R} \right)^2 \sum_{n=-\infty}^{\infty} J_n^2(\Delta) \cdot$$

$$\text{sinc}^2\left[L\left(\frac{x_s}{\lambda R} - nf_1 - \frac{\sin\theta_i}{\lambda} \right) \right] \sin J_n^2\left(\frac{Ly_s}{\lambda R} \right) \quad (B.13)$$

This result is discussed at the end of Sec. 3.2. $I(x_s,y_s)$ is power per unit area. In practice, power measurements are made over fixed apertures. A measurement of grating efficiency, for example, would involve opening the detector aperture so that it accepted all the power in each order of interest. Equation (B.13) may be converted from intensity to power by integrating over x_s,y_s one order at a time. We have already assumed that the order width is small compared to order spacing so all variables, except the sinc arguments, can be regarded as constant over the tight region where the sinc functions are appreciably greater than zero. Using π as the area under sinc^2 and remembering to change variables in the integral gives:

$$P_n = \left(\frac{E_0^2 L^2}{2\eta}\right) J_n^2(\Delta) \cos^2 \theta_{sn} \tag{B.14}$$

Recognizing $(E_0 L)^2/2\eta$ as the incident power P_i, and expanding $J_1(\Delta)$ for small Δ gives an expression for the first-order grating efficiency.

$$P_1/P_i = \left[\frac{\Delta}{2} \cos \theta_{s1}\right] = [\tfrac{1}{2}ka \cos \theta_{s1}(\cos \theta_i + \cos \theta_{s1}]^2 \tag{B.15}$$

for $J_1(\Delta) \simeq \Delta/2$.

At small angles this reduces to the result referred to in Sec. 1.2

$$P_1/P_i \simeq (ka)^2 \tag{B.16}$$

The first-order vector perturbation efficiencies for both s- and p-polarized light reduce to the same small-angle expression.

The effect of two parallel gratings on the surface can be analyzed in a similar fashion. Substitute

$$z(x) = a_1 \sin (2\pi f_1 x + \alpha_1) + a_2 \sin (2\pi f_2 x + \alpha_2) \tag{B.17}$$

into Eq. (B.1) to get the phase delay. The summation in Eq. (B.6) is now a double summation over the product of two Bessel series.

$$E_a(x, y, 0) = E_0[\text{rect } (x/L) \text{ rect } (y/L) \sum_{n=-\infty}^{\infty} \sum_{m=-\infty}^{\infty} J_n(\Delta_1) J_m(\Delta_2)$$

$$e^{j(kx \sin \theta_i + n2\pi f_1 x + m2\pi f_2 x + n\alpha_1 + m\alpha_2)} \tag{B.18}$$

After convolution the sinc arguments are rearranged in a fashion similar to Eq. (B.12) and the intensity is found as

$$I(x_s, y_s) = \frac{1}{2\eta}\left(\frac{E_0 L^2 \cos_s \theta}{\lambda R}\right)^2 \sum_{n=-\infty}^{\infty} \sum_{m=-\infty}^{\infty} J_n^2(\Delta_1) J_n^2(\Delta_2)$$

$$\text{sinc}\left[\frac{L}{\lambda}(\sin\theta_s - \sin\theta_i - n\lambda f_1 - m\lambda f_2)\right] \text{sinc}\left[\frac{Ly_s}{\lambda R}\right] \qquad \text{(B.19)}$$

The diffracted orders are still confined to the incident plane, but there are more of them. Notice that for either n or $m = 0$ the result is essentially that of Eq. (B.13) because $J_0(\Delta) \simeq 1$ for small Δ. Thus the f_1 and f_2 spectrums are both present, essentially undisturbed, as predicted by Eq. (B.13). In addition, the sum and difference frequencies are present as cross terms with amplitudes that depend on both $J_n(\Delta_1)$ and $J_m(\Delta_2)$. The situation is easiest to picture if the two frequencies are well separated and are illustrated in Fig. B.3 for $f_1 \gg f_2$. These types of effects are important when nonsinusoidal grating surfaces, such as those generated by precision machining, are analyzed. If f_1 and f_2 are harmonically related, as will be the case with a nonsinusoidal grating made up of several Fourier components, the cross product terms fall on top of higher-order terms that are already present. For example, if $2f_2 = f_1$, the sum frequency $f_1 + f_2$ is just $3f_2$. However, if an unwanted nonharmonically related frequency is present in the surface, the cross terms can be used to provide clues as to the value of the second fundamental. Knowing the second fundamental frequency may allow it to be eliminated (see Sec. 4.3).

Another case of interest is diffraction from crossed sinusoidal gratings. The analysis is very similar to that of the parallel gratings, except that the f_1 and f_2 spectrums are now located along the x_s and y_s axes and the cross-product terms are off-axis forming rectangles in frequency space. The relationship is given in Chap. 3 as Eq. (3.38).

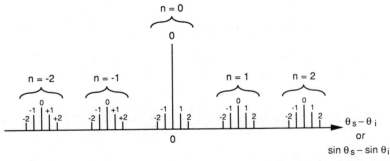

Figure B.3 Diffraction locations for two parallel sinusoidal gratings with $f_1 \gg f_2$.

BSDF Data

The text has gone into considerable detail on the analysis and measurement of BSDF data. The importance of using a data set that really matches a specific should be very clear. There are times when knowing general BSDF levels can be of considerable help. For example, if you have determined by way of analysis that a ZnSe window will work in your application if its mid-IR scatter can be kept below $10^{-3}\ sr^{-1}$ at 30 degrees from specular, then knowing that windows are routinely made at $10^{-5}\ sr^{-1}$ in this region lets you proceed with your design. If you needed $10^{-6}\ sr^{-1}$ then the situation would be a little more difficult. You might actually want your component measured, or you might want to change your design.

Another problem with practical use of BSDF data sets is categorizing the variables associated with the sample, the measuring instrument, and the laboratory reporting methods. An ASTM standard for taking and reporting BRDF (not BTDF) data is expected in the early 1990s. It goes a long way toward establishing a rather flexible standard for reporting data. Thus labs that can convert to and from the standard will be able to compare and/or trade BSDF data. It is expected that any group serious about this field of metrology will comply with the standard. It is expected that a relational database, probably PC-based, will become available that will allow a fast search of hundreds of files to obtain data that fit a particular requirement.

The 32 BSDF scans in this appendix are offered in the spirit of giving only a feeling for values measured from a few samples. Hopefully it will be useful in the short term. Once an accessible national database is available, there will be little need for the microscopic, difficult-to-use, windows on BSDF data provided by the data sets similar to this appendix.

The following chart provides a means of locating the various BSDF data files in the book. The materials measured are listed alphabetically and the corresponding figure numbers given by wavelength region. The data sets of this appendix are given in ascending order of wavelength. This is far short of the relational database just discussed. As you use the list, you may begin to appreciate the value of establishing a national database.

	Visible 0.4 → 0.7 mm (Figure number)	Near-IR 0.7 → 2 mm (Figure number)	Mid-IR 2 → 12 mm (Figure number)
Aluminum, polished	C.6	C.6	C.6
Aluminum on glass	9.1, 9.5		
Beryllium	C.7, 7.4, 7.5, 7.6	C.7, 7.6	C.7, 7.4, 7.5, 7.6
Cloth, black corduroy	7.10		
Copper	C.4		
Dielectric-coated laser mirror	9.11		
Dielectric-coated RLG mirror	6.17c		
Gallium arsenide			C.9
Gallium phosphide	C.3		
Germanium			C.9
Glass, AR coat	C.2		
Glass, uncoated	5.4		
Martin black	7.10		
Molybdenum	C.5, C.8, 4.1, 6.2, 7.7	C.8	C.5, C.8
Nickel on copper #2	C.4		
Nickel on copper #1	4.8		
Paper	7.9, 7.10, 9.15		
Pelical	C.2		
Plexiglass	C.2		
Polyurethane enamel, white	7.9		
Potassium chloride			C.9
Replicated mirrors	9.2		
Silicon carbide	C.5		C.5
Spectralon	C.1, 7.9, 7.10, 7.11		
Steel, ball bearings	9.19		
Sunglasses	C.2		
Tellurium dioxide	C.3	6.10	
White flat spray paint	7.9		
Zinc selinide	8.10		C.9
Zinc on steel		9.17	9.16

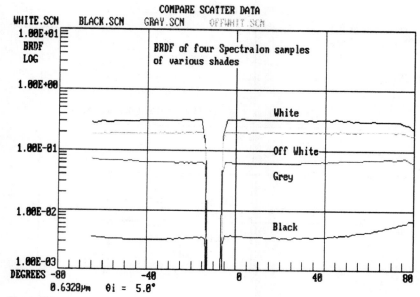

COMPARE SCATTER DATA

WHITE.SCN   BLACK.SCN   GRAY.SCN   OFFWHIT.SCN

BRDF of four Spectralon samples of various shades

White
Off White
Grey
Black

0.6328μm   θi = 5.0°

Figure C.1

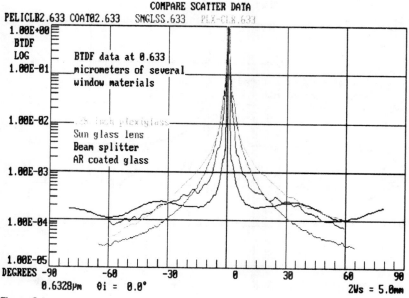

COMPARE SCATTER DATA

PELICLB2.633 COAT02.633   SNGLSS.633   PLX-CLR.633

BTDF data at 0.633 micrometers of several window materials

.25 inch plexiglass
Sun glass lens
Beam splitter
AR coated glass

0.6328μm   θi = 0.0°   2Ws = 5.0mm

Figure C.2

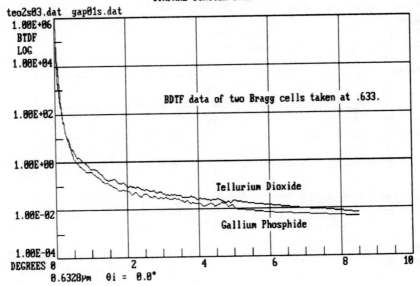

Figure C.3

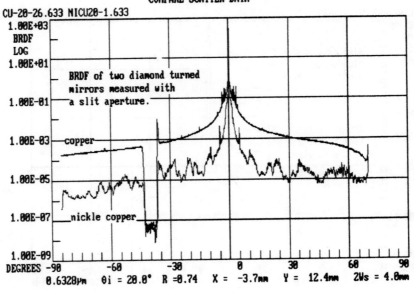

Figure C.4

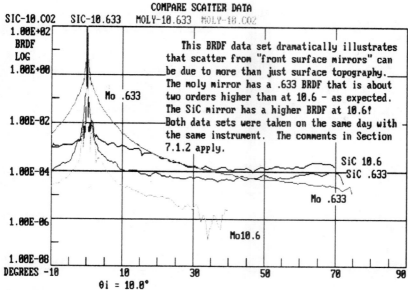

COMPARE SCATTER DATA

SIC-10.CO2   SIC-10.633   MOLY-10.633   MOLY-10.CO2

This BRDF data set dramatically illustrates that scatter from "front surface mirrors" can be due to more than just surface topography. The moly mirror has a .633 BRDF that is about two orders higher than at 10.6 – as expected. The SiC mirror has a higher BRDF at 10.6! Both data sets were taken on the same day with the same instrument. The comments in Section 7.1.2 apply.

θi = 10.0°

Figure C.5

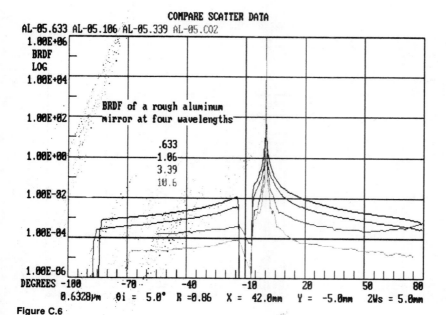

COMPARE SCATTER DATA

AL-05.633 AL-05.106 AL-05.339 AL-05.CO2

BRDF of a rough aluminum mirror at four wavelengths

.633
1.06
3.39
10.6

0.6328μm   θi = 5.0°   R =0.86   X = 42.0mm   Y = -5.0mm   2Ws = 5.0mm

Figure C.6

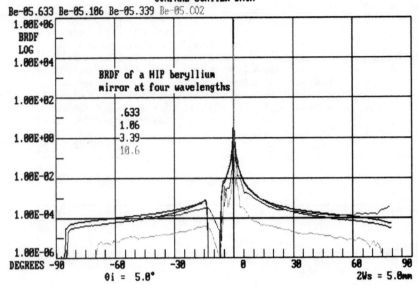

Figure C.7

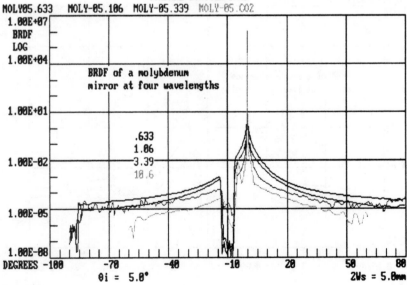

Figure C.8

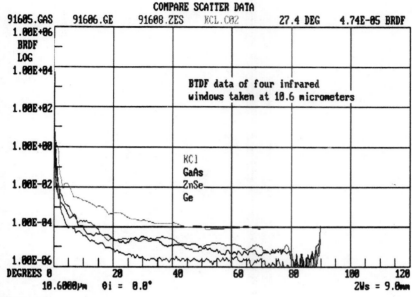

Figure C.9

Bibliography

ANSI/ASME, "Surface Texture (Surface Roughness, Waviness, and Lay)," ANSI/ASME Standard no. B46.1-1985, Am. Soc. Mech. Eng., New York, 1985.

Azzam, R. M. A., and N. M. Bashara, *Ellipsometry and Polarized Light*, North Holland, New York, 1977.

Bamberg, J., "Stray Light Analysis with the HP-41C/CV Calculator," *Proc. SPIE*, 384 (1983).

Barrick, D. E., *Radar Cross Section Handbook*, Plenum, New York, 1970, chap. 9.

Beckmann, P., and A. Spizzichino, *The Scattering of Electromagnetic Waves from Rough Surfaces*, Pergamon, New York, 1963.

Bell, B. W., and W. S. Bickel, "Single Fiber Light Scattering Matrix: An Experimental Determination," *Appl. Opt.*, vol. 20, no. 22, 1981, p. 3874.

Bendat, J. S., and A. G. Piersol, *Random Data: Analysis and Measurement Procedures*, Wiley-Interscience, New York, 1971.

Bendat, J. S., and A. G. Piersol, *Engineering Applications of Correlation and Spectral Analysis*, 2d ed., Wiley, New York, 1986.

Bennett, H. E., and J. O. Porteus, "Relation between Surface Roughness and Specular Reflectance at Normal Incidence," *J. Opt. Soc. Am.*, vol. 51, 1961, p. 123.

Bennett, J. M., and L. Mattsson, *Introduction to Surface Roughness and Scattering*, Opt. Soc. of Am., Washington D.C., 1989.

Bickel W. S., J. F. Davidson, D. R. Huffman, and R. Kilkson, "Application of Polarization Effects in Light Scattering: A New Biophysical Tool," *Proc. Nat. Acad. Sci.*, vol. 73, no. 2, 1976, p. 488.

Bickel W. S., V. Iafelice, and G. Videen, "The Role of Polarization in the Measurement and Characterization of Scattering," *Proc. SPIE*, 679 (1986), p. 91.

Bickel W. S., R. R. Zito, and V. Iafelice, "Polarized Light Scattering from Metal Surfaces," *J. Appl. Phys.*, vol. 61, no. 12, 1987, p. 5392.

Bjork, D. R., K. A. Klicker, and F. M. Cady, "Predicting Laser Port Scatter," *Proc. SPIE*, 967–07 (August 1988).

Bohren, C. E., and D. R. Huffman, *Absorption and Scattering of Light by Small Particles*, Wiley, New York, 1983.

Breault, R. P., "Current Technology of Stray Light," *Proc. SPIE*, 675–01 (1986).

Breault, R. P., and A. W. Greynolds, "APART/PADE 8.6 A Deterministic Computer Program Used to Calculate Scattered and Diffracted Engery," *Proc. SPIE*, 675–22 (1986).

Brown, N. J., "Optical Fabrication," Lawrence Livermore National Laboratories, Misc. 4476 Rev. 1, August, 1989.

Burnham, M., "The Mechanics of Micromachining," *Proc. SPIE*, 93–06 (1976).

Cady, F. M., D. R. Cheever, K. A. Klicker, and J. C. Stover, "Comparison of Scatter Data From Various Beam Dumps," *Proc. SPIE*, 818–21 (1987).

Cady, F. M., J. C. Stover, T. F. Schiff, K. A. Klicker, and D. R. Bjork, "Measurement of Very Near Specular Scatter," *Proc. SPIE*, 967–26 (1988).

Cady F. M., D. R. Bjork, J. Rifkin, and J. C. Stover, "BRDF Error Analysis," *Proc. SPIE*, 1165–13 (1989).

Cady F. M., D. R. Bjork, J. Rifkin, and J. C. Stover, "Linearity in BSDF Measurement," *Proc. SPIE*, 1165–44 (1989).

Cheever, D. R., F. M. Cady, K. A. Klicker, and J. C. Stover. "Design Review of a Unique Complete Angle-Scatter Instrument (CASI)," *Proc. SPIE*, 818, "Current Developments in Optical Engineering II," (1987), p. 13.

Church, E. L., and J. M. Zavada, "Residual Surface Roughness of Diamond-Turned Optics," *Appl. Opt.*, vol. 14, 1975, 1788.

Church. E. L., H. A. Jenkinsons, and J. M. Zavada, "Measurement of the Finish of Diamond-Turned Metal Surfaces by Differential Light Scattering," *Opt. Eng.*, vol. 16, 1977, p. 360.

Church, E. L., "Corrections to Stylus Measurements of Surface Finish," *J. Opt. Soc. Am.*, vol. 68, 1978, 1425A–1426A.

Church, E. L., H. A. Jenkinson, and J. M. Zavada, "Relationship between Surface Scattering and Microtopographic Features," *Opt. Eng.*, vol. 18, no. 2, 1979, p. 125.

Church, E. L. "Interpretation of High-Resolution X-Ray Scattering," *Proc. SPIE*, 257–254 (1980). (Numerical results for a $1/f$ spectrum contain an error.)

Church, E. L., M. R. Howells, and T. V. Vorburger, "Spectral Analysis of the Finish of Diamond Turned Mirror Surfaces," *Proc. SPIE*, 315–34 (1982).

Church, E. L., and H. C. Berry, "Spectral Analysis of the Finish of Polished Optical Surfaces," *Wear*, vol. 83, 1982, p. 189–201.

Church, E. L., P. Z. Takacs, "Statistical and Signal Processing Concepts in Surface Metrology," *Proc. SPIE*, 645–107 (1986).

Church, E. L., "Models For the Finish of Precision Machined Optical Surfaces," *Proc. SPIE*, 676–23 (1986).

Church, E. L., G. M. Sanger, and P. Z. Takacs, "Comparison of WYKO and TIS Measurements of Surface Finish," *Proc. SPIE*, 749–11 (1987).

Church, E. L., "Comments on the Correlation Length," *Proc. SPIE*, 680–18 (1987).

Church, E. L., "Fractal Surface Finish," *Appl. Opt.*, vol. 27, no. 8, April, 1988.

Church, E. L., and P. Z. Takacs, "Instrumental Effects in Surface Finish Measurement," *Proc. SPIE*, 1009–04 (1988).

Church, E. L., J. C. Dainty, D. M. Gale, and P. Z. Takacs, "Comparison of Optical and Mechanical Measurements of Surface Finish," *Proc. SPIE*, 954–27 (1988).

Church, E. L., Private Communication: This useful relationship was first discovered serendipitously while checking Q values with a calculator and then later confirmed by derivation (1989).

Church, E. L., and P. Z. Takacs, "Subsurface and Volume Scattering From Smooth Surface,." *Proc. SPIE*, 1165–04 (1989).

Church, E. L., P. Z. Takacs, and T. A. Leonard, "The Prediction of BRDFs from Surface Profile Measurements," *Proc. SPIE*, 1165–10 (1989).

Dagnall, H., *Exploring Surface Texture*, Rank Taylor Hobson, Leicester, England, 1980.

Davies, H., *Proc. Inst. Elec. Engrs.*, vol. 101, 1954, p. 209.

DeWitt, D. P., "Inferring Temperature from Optical Radiation Measurements," *Opt. Eng.*, vol. 24, no. 4, 1986, p. 596.

DeWitt, D. P., and G. D. Nutter (eds.), *Theory and Practice of Radiation Thermometry*, Wiley-Interscience, New York, 1989, p. 1136.

Dolan, A., "An Interactive Graphical, CAD Integrated Tool for Stray Radiation Analysis," *Proc. SPIE*, 675–18 (1986).

Elson, J. M., and J. M. Bennett, *Opt. Eng.*, vol. 18, 1979, p. 116.

Foo, L. D., "Computer Analysis of Background Radiation Sources for a Staring IRCCD Camera," M.S. thesis, University of Arizona, Tucson, 1985.

Freniere, E. R., "Simulation of Stray Light in Optical Systems with the GUERAPIII," *Proc. SPIE*, 257–78 (1980).

Goodman, J. W., *Introduction to Fourier Optics*, McGraw-Hill, New York, 1968.

Greynolds, A., "Formulas for Estimating Stray-Radiation Levels in Well-Baffled Optical Systems," *Proc. SPIE*, 257 (1980).

Gu, Z. H., R. S. Dummer, A. A. Maradudin, and A. R. McGurn, "Experimental Study of the Opposition Effect in the Scattering of Light From a Randomly Rough Metal Surface," *Appl. Opt.*, vol. 28, no. 3, 1989, p. 537.

Gu, Z. H., R. S. Dummer, A. A. Maradudin, and A. R. McGurn, "Opposition Effect in the

Scattering of Light from a Random Rough Metal Surface," *Proc. SPIE*, 1165–05 (1989).

Hancock, J. C., *Principles of Communication Theory*, McGraw-Hill, New York, 1961.

Harvey, J. E., *Light Scattering Characteristics of Optical Surfaces*, Ph.D. dissertation, University of Arizona, Tucson, 1976.

Harvey, J. E., "Surface Scatter Phenomena: a Linear, Shift-Invariant Process," *Proc. SPIE*, 1165–42 (1989).

Hill, D. P., R. L. Shoemaker, D. P. De Witt, D. R. Gaskell, T. F. Schiff, J. C. Stover, D. White, and K. M. Gaskey, "Relating Surface Scattering Characteristics to Emissivity Changes During the Galvanneal Process," *Proc. SPIE*, 1165–7 (1989).

Hunt, A. J., and D. R . Huffman, "A New Polarization-Modulated Light Scattering Instrument," *Rev. Sci. Instrum.*, vol. 44, no. 12, 1973, p. 1753.

Huntley, W. H., "Grating Interferometers," Lasers 80, T Japan, 1980.

Iafelice, V. J., and W. S. Bickel, "Polarized Light-Scattering Matrix Elements for Select Perfect and Perturbed Optical Surfaces," *Appl. Opt.*, vol. 26, no. 12, 1987, p. 2410.

Iizuka, K., *Engineering Optics*, Springer-Verlag, Berlin, 1985.

Ishimaru, A., *Wave Propagation in Random Media*, Academic Press, New York, 1978.

Jenkins, F. A., and H. E. White, *Fundamantals of Optics*, McGraw-Hill, New York, 1976.

Jenkins, G. M., and D. G. Watts, *Spectral Analysis and Its Applications*, Holden-Day, San Francisco, 1968.

Keller, J. B., "Geometrical Theory of Diffraction," *J. Opt. Soc. Am.*, vol. 52, 1962, p. 116.

Klicker, K. A., J. C. Stover, D. R. Cheever, and F. M. Cady, "Practical Reduction of Instrument Signature in Near Specular Light Scatter Measurements," *Proc. SPIE*, 818–26 (1987).

Klicker, K. A., and D. R. Bjork, "Model of Port Scatter from Lasers," Final Report to U.S. Army White Sands Missile Range, Contract No. DAAD07-87-0083, 1988.

Klicker, K. A., J. C. Stover, and D. J. Wilson, "Near Specular Measurement Techniques for Curved Samples," *Proc. SPIE*, 967 (1988).

Krauss, H. L., C. W. Bostian, and F. R. Raab, *Solid State Radio Engineering*, Wiley, New York, 1980.

Leader, J. C., "Analysis and Prediction of Laser Scattering from Rough Surface Materials," *JOSA*, vol. 69, 1979, p. 610–628.

Lee, W. W., L. M. Scherr, and M. K. Barsh, "Stray Light Analysis and Suppression in Small Angle BRDF/BTDF Measurement," *Proc. SPIE*, 673–32 (1986), p. 207.

Leonard, T. A., M. A. Pantoliano, and J. Reilly, "Results of a CO_2 BRDF round robin," *Proc. SPIE*, 1165–12 (1989).

Leonard, T. A., and M. A. Pantoliano, "BRDF round robin," *Proc. SPIE*, 967–22 (1988).

Likeness, B. K., "Stray Light Simulation with Advanced Monte Carlo Techniques," *Proc. SPIE*, 107–80 (1977).

McGary, D. E., J. C. Stover, J. Rifkin, F. M. Cady, and D. R. Cheever, "Separation of Bulk and Surface Scatter from Transmissive Optics," *Proc. SPIE*, 967 (1988).

McGillem, C. D., and G. R. Cooper, *Continuous and Discrete Signal and System Analysis*, Holt, New York, 1984.

McNeil, J. R., W. C. Herrman, and J. C. Stover, "Light Scattering Characteristics of Some Metal Surfaces—A Smoothing Effect?" *Proc. Laser Damage Symposium*, Boulder, Colo., 1983.

Maradudin, A. A., and D. L. Mills, *Phys. Rev.*, vol. B11, 1975, p. 1392.

Marvin, A., F. Toigo, and V. Celli, *Phys. Rev.*, vol. B11, 1975, p. 2777.

Mathis, R. C., "A Lunar Echo Study at 425 mcs," Ph.D. dissertation, University of Texas, 1963.

Maxwell, J. R., J. Beard, S. Weiner, D. Ladd, and S. Ladd, "Bidirectional Reflectance Model Validation and Utilization," Technical Report AFAL-TR-73-303, Wright-Patterson AFB, 1973.

Nicodemus, F. E., J. C. Richmond, J. J. Hsia, I. W. Ginsberg, and T. Limperis, *Geometric Considerations and Nomenclature for Reflectance*, U.S. Dept. of Commerce, Washington, D.C., NBS Monograph 160, 1977.

Noll, R. J., and P. E. Glenn, "Optical surface analysis code (OSAC)," *Proc. SPIE*, 362–15 (1982).

Orazio, F. D., W. K. Stowell, and R. M. Silva, "Instrumentation of a Variable Angle Scatterometer (VAS)," *Proc. SPIE*, 362–28 (1982).

Paumi, J. D., "Laser vs. Camera Inspection in the Paper Industry," *Tappi Journal*, November, 1988.

Rayleigh, Lord. *Proc. R. Soc.*, A79–399 (1907).

Rice, S. O., *Commun. Pure Appl. Math.*, vol. 4, 1951, p. 351.

Rifkin, J., "Design Review of a Complete Angle Scatter Instrument," *Proc. SPIE*, 1036–15 (1988).

Rifkin, J., D. E. McGary, K. H. Kirchner, and D. J. Wilson, "Raster Area Scatter Measurements and Sample Uniformity," *Proc. SPIE*, 967 (1988).

Rock, D., "ORDAS—A New Ray Based Stray Radiation Analysis Program," *Proc. SPIE*, 675–19 (1986)

St. Clair Dinger, A., "An Interactive Program for the Computation of Stay Radiation in Infrared Telescopes," *Proc. SPIE*, 675–21 (1986)

Schiff, T. F., J. C. Stover, D. R. Cheever, and D. R. Bjork, "Maximum and Minimum Limitations Imposed on BSDF Measurements," *Proc. SPIE*, 967–06 (1988).

Schiff, T. F., and J. C. Stover, "Surface Statistics Determined From IR Scatter," *Proc. SPIE*, 1165–06 (1989).

Shurcliff, W. A., *Polarized Light*, Harvard University Press, Cambridge, Mass., 1962.

Siegman, A. E., *Lasers*, University Science Books, Mill Valley, Calif., 1986.

Squires, G. L., *Practical Physics*, Cambridge University Press, Cambridge, England, 1985.

Stover, J. C., "Roughness Measurement by Light Scattering," in A. J. Glass and A. H. Guenther (eds), *Laser Induced Damage in Optical Materials*, U.S. Government Printing Office, Washington, D.C., 1974, p. 163.

Stover, J. C., "Roughness Characterization of Smooth Machined Surfaces by Light Scattering," *Appl. Opt.*, vol. 14, no. 8, 1975, p. 1796.

Stover, J. C., "Spectral Density Function Gives Surface Roughness," *Laser Focus*, February, 1976a, p. 83.

Stover, J. C., "Surface Characteristics of Machined Optics," *Proc. SPIE*, 93, "Advances in Precision Machining of Optics," (1976b).

Stover, J. C., and C. H. Gillespie, "Design Review of Three Reflectance Scatterometers," *Proc. SPIE*, 362–29, "Scattering in Optical Materials," (1982), p. 172.

Stover, J. C., S. A. Serati, and C. H. Gillespie, "Calculation of Surface Statistics from Light Scatter," *Opt. Eng.*, vol. 23, no. 4, 1984, p. 406.

Stover, J. C., and B. Hourmand, "Comparison of Roughness Measurements by Differential Scatter and TIS," *Proc. SPIE*, 511–01, "Stray Radiation IV," (1984), p. 2.

Stover, J. C., and B. Hourmand, "Some Deviations Associated with the Vector Perturbation Theory," *Proc. SPIE*, 511–04, "Stray Radiation IV," (1984), p. 12.

Stover, J. C., F. M. Cady, and E. Sklar, "Measurement of Low Angle Scatter," *Opt. Eng.*, vol. 24, no. 3, 1985, p. 404.

Stover, J. C., "Near Specular Light Scatter Measurements," *Proc. Lasers 87*, 1987a.

Stover, J. C., "Overview of Current Scatterometer Measurements and the Impact on Optical Systems," *Proc. SPIE*, 776–07 (1987b).

Stover, J. C., K. A. Klicker, D. R. Cheever, and F. M. Cady, "Reduction of Instrument Signature in Near Angle Scatter Measurements," *Proc. SPIE*, 749, "Metrology: Figure and Finish," (1987), p. 46.

Stover, J. C., C. H. Gillespie, F. M. Cady, D. R. Cheever, K. A. Klicker, "Wavelength Scaling of BRDF Scatter Data," *Proc. SPIE*, 818, "Current Developments in Optical Engineering II," (1987), p. 62.

Stover, J. C., C. H. Gillespie, F. M. Cady, D. R. Cheever, and K. A. Klicker, "Comparison of BRDF Data from Two Scatterometers," *Proc. SPIE*, 818, "Current Developments in Optical Engineering II," (1987), p. 68.

Stover, J. C., and J. Rifkin, "Analysis of Process Induced Damage by Subsurface Scatter Measurement," Final Report to RADC/ESM, Hanscom AFB, Contract No. F19628-C-0151.

Stover, J. C., J. Rifkin, D. R. Cheever, K. H. Kirchner, T. F. Schiff, "Comparisons of Wavelength Scaling Predictions to Experiment," *Proc. SPIE*, 967–05 (1988).

Stover, J. C., "Optical Scatter Measurements and Specifications," *Lasers and Optronics*, vol. 7, no. 8, August, 1988.

Stover, J. C., "Scatter from Optical Components: An Overview," *SPIE Proc.*, 1165–01 (1989).

Stover, J. C., M. L. Bernt, D. E. McGary, and J. Rifkin, "Investigation of Anomalous Scatter from Beryllium Mirrors," *Proc. SPIE*, 1165–43 (1989).

Takacs, P. Z., R. C. Hewitt, and E. L. Church, "Correlation between the Performance and Metrology of Glancing Incidence Mirrors Containing Millimeter Wavelength Shape Errors," *Proc. SPIE*, 749–119–124 (1987).

Tanka, F., and D. P. DeWitt, "Theory of a New Radiation Thermometry Method and Experimental Study Using Galvannealed Steel Specimens," accept. for pub. in *Trans. Soc. Inst. & Cont. Eng.*, Japan, 1989.

Thomas, T. R., *Rough Surfaces*, Longman, New York, 1982.

van de Hulst, H. C., *Light Scattering by Small Particles*, Wiley, New York (1957).

Verdeyen, J. T., *Laser Electronics*, 2d ed., Prentice-Hall, Englewood Cliffs, N.J., 1989.

Vernold, C. L., "Application and Verification of Wavelength Scaling for Near Specular Scatter Predictions," *Proc. SPIR*, 1165–03 (1989).

Wang, Y., "Comparison of BRDF Theories with Experiment," Ph.D. dissertation, University of Arizona, Tucson, 1983.

Wilson, S. R., G. A. Al-Jumaily, and J. R. McNeil, "Nonlinear Characteristics of a Stylus Profilometer," *Proc. SPIE*, 818–10 (1987).

Wolf, E., and E. W. Marchand, "Comparison of the Kirchhoff and the Rayleigh-Sommerfield Theories of Diffraction at an Aperture," *J. Opt. Soc. Am.*, vol. 54, 1964, p. 587.

Wolfe, W. L., and Y. Wang, "Comparisons of Theory and Experiment for BRDF of Microrough Surfaces," *Proc. SPIE*, 362–08 (1982).

Yariv, A., *Introduction to Optical Electronics*, 2d ed., Holt, New York, 1976.

Young, M., "Objective Measurement and Characterization of Scratch Standards," *Proc. SPIE*, 362–17 (August 1982).

Young, R. P., "Mirror Scatter Measurement Facility Comparison." AEDC-TR-75-68 September, 1975.

Young, R. P., "Degradation of Low Scatter Mirrors by Particulate Contamination," *Opt. Eng.*, vol. 15, no. 6, November–December 1976a, pp. 516–520.

Young, R. P., "Degradation of Mirror BRDF by Particulate Contamination," AEDC-TR-177 [or AD-B015792] December, 1976b.

Zito, R. R., and W. S. Bickel, "Light Scattering from Twisted Metal Cylinders," *Appl. Opt.*, vol. 25, no. 11, 1986, p. 1833.

Index

About the Author

John C. Stover is Vice President of Engineering at Toomay, Mathis & Associates, Inc. A specialist in light scatter metrology, he has held a variety of academic and industrial positions including Associate Professor of Electrical Engineering at Montana State University, Senior Research Engineer at Rockwell International, and self-employed consultant in the field of light scatter. He has published numerous technical papers on optical scattering and electro-optics and is a recognized leader in the field.